ROCKY MOUNTAIN BERRY BOOK

BOB KRUMM

FALCON™
Helena, Montana

Acknowledgments

There are a host of people that helped make this book possible. Mike Aderhold, Public Information Officer for the Montana Fish, Wildlife, and Parks in Kalispell, offered invaluable assistance. He helped to edit the book and told me of people to contact for recipes and information.

Bob Giurgevich helped immensely with editing and botanical advice. Earl Jensen provided botanical editing and several photos. Dr. Dee Strickler, Columbia Falls, Montana, helped with botanical editing and photos as well.

Betty Bindl, Trout Creek, Montana, provided me with her wonderful book of huckleberry recipes and lots of information about the Trout Creek Huckleberry Festival.

Charlotte Heron, Missoula, Montana, supplied me with lots of unique wild berry recipes.

Theo Hugs, who works for Bighorn Canyon National Recreation Area at Ft. Smith, Montana, gave me several Crow Indian recipes and lots of bits of natural history and Indian history.

Eternal thanks to my parents, Donald and Emily, for giving me an outdoor education from early childhood.

Sincere appreciation goes to my friend, Dot Heggie, who urged me to write the *Rocky Mountain Berry Book*. Her coaxing and encouraging helped to make the dream a reality.

Finally, thanks to all the folks who submitted recipes, gave advice, cheered me along, and offered encouragement.

Copyright © 1991 by Falcon Press Publishing Co., Inc.,
Helena and Billings, Montana

Design, typesetting, and other prepress work by Falcon Press, Helena, Montana.
Photography by Bob Krumm, unless otherwise indicated.
Glossary illustrations by Earl Jensen

Library of Congress Number: 90-81720

ISBN 1-56044-040-6

Contents

Foreword

Most people in the West have one particular passion that sustains them in the outdoors.

For some, it's the vision of fly fishing for trout on a foam-flecked river flowing between high mountain peaks.

For others, it's the experience of carrying a backpack through an alpine meadow sprinkled with the colors of countless wildflowers.

Still others are in their glory only when the aspens turn gold, the mornings turn frosty cold, and the sounds of bugling bull elk are in the air.

But there's no finer compliment that can be paid an outdoorsman anywhere than to be called a person for all seasons. And Bob Krumm is one of these.

For that reason, it's fitting he should write a book on the wild fruits and berries of the Rockies.

But he could just as easily write one on gathering wild asparagus or picking morel mushrooms, or trout fishing on big rivers or small, or hunting big or small game, or ice fishing, or cross-country skiing, or hiking, or outdoor photography, or wildflowers, or even doing battle with Rocky Mountain grasshoppers to preserve the vegetables in his garden.

All of these things have become topics for his column, Wyoming Outdoors, that he mails in from his home in Sheridan each week for the Billings Gazette's outdoor section.

As his editor for that column, it's always easy to see that what rings true most often, no matter what the topic, is the love he has for the varied outdoor life.

From his early years growing up in Michigan to his adult life spent in Wyoming, he's been a student of the outdoors striving to learn all he could about the world around him.

As a result, each season brings him something new to learn and special to do in the outdoors.

Each day in the field presents new challenges for him and new things to see.

And, he's never happier than when he can discover some new, previously unknown treasure—whether that's finding a new place to pick the fruits of the wild, exploring the distant reaches of a new trout stream, or simply spending a day out among the birds and animals that call the Rockies home.

His *Rocky Mountain Berry Book* is just one stop along the road of the outdoorsman's life that shapes Bob Krumm—but it's a good stop for this man for all seasons.

Mark Henckel, Outdoor Editor
The Billings Gazette
Billings, Montana.

Introduction

Wild berries and fruits have been a part of my experience almost from my first memories. I can remember my parents taking my brother, sister, and me for picnics on Sunday afternoons. We sat on the bank of an old, winding river and had chicken and potato salad. After lunch, Dad would go fishing for awhile. We kids might fish, too, but would soon tire of it.

Mom would try to think of other things for us to do. Many times she would lead us to the nearest berry patch. We would pick berries for awhile and maybe pick enough so mom could make a pie when we got home.

Other times, we went out specifically for berries for jam and jelly—no fishing, just picking. We would get enough berries for a batch or two of jelly. Those wild berry jams and jellies coupled with the domestic berry and fruit jams and jellies filled our root cellar and provided sweets for a long winter.

As I grew older, I enjoyed learning about the edible fruits and berries of an area. I liked to figure out where the berries would grow and where the juiciest, tastiest ones would be. I got so I could recognize an edible wild berry or fruit in the winter or spring by the characteristics of the bush—its height, shape, color, branching, thorns, leaves, flowers, and location. By knowing where a particular berry was growing, I could make a mental list and swing by when I thought it would be ripe.

This mental inventory served me well—particularly when my usual berry patches didn't have good crops. It seemed that I always knew enough reserve berry patches that I could end up getting a bucketful of berries even in lean years.

I have tried to design this book so that you can recognize the berries at various times of the year. That way you too can do some preseason scouting and have several berry patches in mind when the berries are ripe.

Over the years I have found that berry picking offers lots of satisfaction and is just as enjoyable an outdoor pastime as fishing or hiking. I often combine berry picking with my hiking or fishing trips and double my satisfaction. Berry picking is something you can do alone, with another person, or with your family.

Using this book as a field guide, you will be able to identify the common, edible wild berries and fruits of the northern Rocky Mountain states, such as Colorado, Wyoming, Montana, Idaho, and Utah. The book will also be good for some of the western parts of the plains such as Nebraska, South Dakota, and North Dakota.

I hope *The Rocky Mountain Berry Book* will help you avoid the problems I had with field guides and cookbooks. I got tired of field guides that identified a plant meticulously, and then, as an addendum, stated the berry or fruit was "edible" but went no further than to say "makes good jellies and wines." Many recipe books give detailed instructions on how to make jellies, pies, muffins, and wines. However, they don't describe the berry well enough for the average person to recognize it—even if the plant wrapped itself around the person's ankle. I think this combination book will solve these problems and allow you to identify and prepare the berry or fruit in some exciting ways.

Berry picking is enjoyable. With a little effort, you can turn the fruits and berries into some great taste treats. Many berries can be eaten as is; others need to be doctored a bit. Whatever the case, there is a wealth of natural, luscious, wild berries and fruits out there. The good Lord created them for us to enjoy, so give them a try. You might find that berry picking will become a pleasant part of your life. You may discover that some of your most rewarding days spent out-of-doors were when you picked buckets of berries and made them into your own taste treats. Enjoy! 🍒

Hazards

Berry picking is a lot of fun and enables you to have a cupboard full of tasty berry products, but you should be aware of hazards in the berry patch. The hazards fall under three categories: animal, plant, and chemical.

In the mountain West there are a number of animals that can spoil a berry-picking trip. In Montana and Wyoming, especially around Yellowstone and Glacier national parks, there are grizzly bears. These big bruins like to eat berries as much or more than you do. They especially like huckleberries.

One precaution you should take in grizzly country involves letting a bear know you're coming by making plenty of noise. Sing, tell stories, wear a bear bell, whistle—making enough noise so that you don't surprise a bear. If you are picking berries and you spot a bear, leave cautiously. Don't run. Grizzlies can't climb trees, so if you can make it to a good-sized tree, you can escape. The National Park Service puts out a good brochure on bear safety that you can pick up at information offices and visitor centers.

Black bears can be just as big a problem as grizzlies. They live in all the forested mountain regions. Again, making plenty of noise is your best precaution against surprising a black bear.

It may come as a surprise to you, but moose and bison can be just as dangerous as bears. Give these critters plenty of leeway.

Rattlesnakes are prevalent throughout the West. You'll find them on the plains and in the canyons and foothills. Your best defense is to watch your step. If you hear a rattlesnake buzz, locate it, then back away from it. Don't kill it; just circle around it and be on your way.

Other animals can be problems. I find that running into a nest of yellow jackets or paper wasps can be worse than running into rattlesnakes. Pesky wasps can sting so quickly that a person with an allergy to them can be in a big world of hurt in a few seconds. I have friends who carry hypodermic syringes with antidote to counteract wasp stings.

Red ants like black currants. The ants will defend the bushes actively. Their bite stings, due to the formic acid they exude. If you come across a currant bush defended by ants, pass it by and look for an undefended one.

Unfortunately, there are a number of poisonous plants that can make a berry-picking trip a real disaster. Tops on the list is nightshade. It grows along stream courses in the foothills and plains, and has purple, star-shaped flowers. The pea-sized fruits look like little Italian tomatoes, elliptical and red. If eaten, they can cause paralysis and death. It's ironic that this poisonous plant is cousin to the tomato and potato, two of our most important food plants.

Another plant that can make your trip tough is baneberry. This perennial grows in moist areas along mountain streams. It has red or white glossy berries. The berries are so glossy that another common name for the plant is chinaberry. The berries look very tasty, but they aren't! The poison in the berry acts on the heart. The berries have caused several deaths in children. Eating six berries is sufficient to increase pulse, induce dizziness, and cause burning in the stomach and cramps.

Poison ivy occurs in the plains, foothills, and canyons. This three-leafed, low-growing vine has an oil that causes severe itching. One of the easiest ways to avoid getting poison ivy is to be able to recognize it and then avoid it. Poison ivy reaches about a foot in height, and has glossy green compound leaves with three leaflets. In the fall, the leaves turn deep red or orange. Poison ivy has small white berries with noticeable seams in them; these berries persist through the winter.

Whenever you are berry picking at lower elevations such as foothills and plains, poison ivy is a definite possibility. Make sure you wear pants, socks, and long sleeved shirts—have as little bare skin exposed as possible. If you think that you might have been exposed to poison ivy, wash the affected area with lots of soap and water as soon as possible.

In general, avoid all white, waxy berries. No white berries are edible in the Rocky Mountain West. Many of the red, waxy berries in the area should be avoided as well.

The third hazard to be aware of is chemicals, namely herbicides. Many commercial timber companies will spray an area after they have logged it to eliminate broad-leafed plants and shrubs and to encourage the establishment of conifers. If the plants you intend to pick are sickly-looking and have yellow or brown leaves, don't pick the berries. Herbicides can be carcinogenic. Remember that many highway departments regularly spray the right-of-ways of state and county roads. Again, if the plant is anemic, discolored, or dying, don't pick the berries.

When making jams and jellies listed in this book, it is wise to preserve them by using the USDA-recommended boiling water bath technique. Use canning jars that require a lid and screw band. Always sterilize the jars and lids by boiling them, and then pour the hot jelly or jam into the hot jars. Wipe the jar rim clean, place the lid on the jar, and tighten the band securely. Place the jars in a kettle of boiling water. There should be an inch of water over the jars. Boil ten minutes for half-pint and pint jars.

One final note of caution: I have tried to thoroughly describe these common wild berries and fruits by word and photos; make sure you read the description and study the photos. If you still aren't sure, don't pick the berry. There are plenty of field guides on the market. Consult another one, or get an expert berry picker to help you identify the edible berries. Once you learn the edible berries and fruits, recognizing them will be much easier.

Do have a safe and enjoyable time picking and cooking the wild berries and fruits that grace the mountain West. 🍒

Buffaloberry

Buffaloberry exemplifies the eastern Rocky Mountain foothills and plains. It is a tough, thorny shrub that survives well where other brushy species might not stand a chance. Its berries provide food for numerous birds and mammals while its spiny branches provide escape cover for ring-necked pheasants, sharp-tailed grouse, Hungarian partridge, and cottontail rabbits.

When I think of buffaloberry, I see the Bighorn River in August near Ft. Smith, Montana, with buffaloberry bushes lining the banks. The berries are so plentiful in my mind's eye that the bushes appear red and the river-banks are red-tinted.

When buffaloberries are plentiful, the picking is easy. Did I say easy? I meant easier. You see, buffaloberries never are easy to pick because the multitude of thorns that adorn the branches make it impossible for a person to pick the berries without getting impaled several times. Believe me, those two to three-inch needles can cause a lot of hurt. Still, buffaloberry is worth the discomfort, for the berries yield some of the most flavorful jelly imaginable.

I'm a relative newcomer to buffaloberry. When I moved to Sheridan, Wyoming, in 1979, I noticed them. Several people told me that the berries made good jelly, but that it was better to wait until frost before picking them. After a frost, a person merely had to place a tarp under the bushes, then shake or beat them. A person could then pick up a bucketful of buffaloberries the easy way.

I thought about the buffaloberries off and on through the years but I didn't pick any until 1988, when a drought had reduced the crop of chokecherries, currants, gooseberries, and wild plums to nearly nothing. Amazingly, there were plenty of buffaloberries. Since the blackbirds were busy harvesting the berries, I didn't think it prudent to wait for a frost, so my son, James, and I went for a bucket of berries in early August.

I learned quickly that I couldn't reach into the bush without looking—the thorns taught me that. I also learned that the easiest way to pick buffaloberries was to surround a clump of berries with my thumb and four fingers, then pull outward. In this manner, I avoided the thorns that were in the clump.

It was slow picking, but in an hour and a half James and I had picked a gallon of berries. We ambled home, washed the berries, then boiled them for a while. I extracted the juice and discovered that the juice was not sparkling red as I imagined it would be, but rather a milky-looking red-orange.

I found out later that buffaloberry contains a compound, saponin, which is a foaming agent. It imparts the milky color to the juice. I wasn't too anxious to make jelly from the juice, but enough people had told me that they had made buffaloberry jelly, so I proceeded.

Just as amazing as the milky juice was the red-orange colored jelly that resulted. There was no soapy tint. It left as soon as I added the sugar to the boiling liquid. The jelly appeared as pretty as crab apple jelly.

There are two species of buffaloberry. Both belong to the genus *Sheperdia*. Both species are dioecious, meaning that a single plant is either male or female, not both sexes as many flowers are. One species, bitter buffaloberry, *S. canadensis*, does not have thorny branches but has very bitter berries. The thorny-branched species, *S. argentea*, goes by the common name of silver buffaloberry, which bears sour but edible berries. It is the species described hereafter.

Buffaloberry favors bottomlands and streambanks. It may attain a height of twenty feet, though eight to ten feet seems to be the average height of the bushes I see along the Bighorn River in southeastern Montana and along the creeks in Sheridan County in northeastern Wyoming. The simple leaves appear silvery-green on both surfaces and are one to two inches in length. The leaf shape is oblong with a rounded tip and an entire margin.

The younger branches have a silvery-gray hue. The older branches and trunk appear dark gray. The bark on older buffaloberry becomes quite veined and somewhat shredded. Thorns are quite evident on the branches.

The blossoms occur late April through May. The tiny, nondescript, brown flowers are less than an eighth of an inch long. The berries mature in late July or early August. They will persist until cold weather. The berries range from split pea-size to pea-size (depending on the bush, moisture, and so on). The berry color is red or golden yellow. The male bushes will not have berries—which explains why you will find some bushes loaded with berries and some that have none.

Most years buffaloberries are abundant enough so that you can pick a gallon in two hours' time. Your chances of getting a gallon of berries are ninety percent.

I have never tried storing buffaloberry other than by refrigerating it. I think that if you can't use the berries immediately, the best thing to do is to render (extract) the juice and freeze it. 🍒

Recipes

BUFFALOBERRY JELLY

16 cups buffaloberries
2 cups water
sugar

Wash and stem buffaloberries. Pick them over and discard unwanted ones. Place in a deep pan with water and bring to a boil, stirring to prevent burning. Boil 15 minutes, drain, mash to get all juice out. Strain through a jelly bag. Measure the juice. Add one cup of sugar for each cup of juice. Blend together, stirring until sugar is dissolved. Bring to a boil and test for jellying. When 2 drops run together off the side of spoon (see "sheet test" in glossary), put into hot sterilized jars, skimming first. Cover with paraffin. Makes eight 8-ounce glasses. Two cups of juice make one 8-ounce glass of jelly. (The USDA recommends the boiling water bath method of preserving jams and jellies.)

Bob Giurgevich, Sheridan, Wyoming

BUFFALOBERRY JELLY

For every 2 quarts of fruit add 1 cup of water and crush in a kettle. Boil slowly for 10 minutes, stirring often. Drain off juice. It will be milky. For each 1 cup of juice, add 1 cup of sugar. Bring to a boil and boil until it jells. It will turn a pale to deep orange when you add the sugar. Pour into sterilized jars, attach lids and rings firmly. Process in a boiling water bath for 10 minutes.

If the fruit is extremely ripe or has been through several frosts, you might want to use the recipe with either Surejell or MCP pectin; before a frost the fruit contains enough pectin to jell by itself.

> *"Buffaloberry jelly was my mother's favorite."*
> *Charlotte Heron, Missoula, Montana*

DRYING BUFFALOBERRIES — NATIVE AMERICAN METHOD

1. Wash berries and remove stems.
2. Put berries in a food grinder or grind on a stone to a mushy consistency, and make soft berries into patties.
3. Place patties on wax paper in the sun.
4. Rotate these every day so they do not mold. Patties should be dry in about a week. If they are brittle and break when bent, they are dry.
5. Store in a jar or can with a lid, in a cool, dry place.

Suggested uses for dried berries:

Syrup:
    3 cups berries
    2 cups water
    2 cups sugar

Soak berries in water until tender. Bring berries to a boil and strain to remove seeds. Add sugar. Refrigerate leftover syrup.

Berry Gravy

Make syrup with berries. Thicken syrup with flour and water mixture. Boil until thick, stirring constantly. Remove from heat and store in refrigerater in clean, covered containers.

Margaret and Charles Butterfield,
Preserving Wyoming's Wild Berries and Fruits,
University of Wyoming Agricultural Extension Service, 1981.

BUFFALO KETCHUP

1 gallon buffaloberry pulp (about 1½ gallons berries)
4 cups vinegar
4 cups sugar
½ teaspoon allspice
2 tablespoons pepper
1 tablespoon nutmeg
1 tablespoon cinnamon

Wash berries and boil until tender, then put through colander. Mix in remaining ingredients and cook until quite thick. Put in jars and seal. This is excellent with cold meats.

Leigh Sherman, Billings, Montana.
From the Billings Gazette Cookbook, *September 25, 1983.*

BUFFALOBERRY JAM

9 cups buffaloberry juice and pulp
12½ cups sugar
2 packages Certo fruit pectin

Examine the berries on the trees after the first hard frost. Do not bother to pick them if dry and shriveled. Pick with gloves to protect your hands or beat berries from the branches with a board or shovel onto large cloths, tarps, or old bedspreads.

Wash berries in cold water, remove all leaves.

Cook berries for 5 to 10 minutes in a large pan with enough water to just cover the berries. As it cooks, mash the berries occasionally with a potato masher. Drain and mash through a jelly-making cone.

Using a large kettle to prevent jam from boiling over, follow a basic pectin recipe for jam, using 9 cups juice, 12½ cups sugar, and 2 packages fruit pectin. Recipe may be cut in half if kettle is too small or if you have less juice.

The jam produced is quite red, with a soft jelled consistency.

While experimenting, other pectin products and basic recipes were tried using other amounts of sugar and juice. The color was not as red, but more orange, with separation of the pulp from the juice.

Mrs. Albert Groskinsky, Sidney, Montana.
From the Billings Gazette Cookbook, *September 25, 1983.*

BUFFALOBERRY CONSERVE

4 cups of ripe berries
1½ cups of water
¼ pound chopped seedless raisins
½ cup chopped nut meats
1 chopped orange, including rind
8 cups sugar

Wash and clean 4 cups of fully ripened buffaloberries. Add to a deep saucepan with 1½ cups of water. Cook until the berries have softened. Stir occasionally. Add ¼ pound of chopped seedless raisins, ½ cup of chopped nut meats (walnuts), 1 chopped orange, and 8 cups of granulated sugar. Mix well and stir constantly. Cook the mixture 20-30 minutes. Then skim off the foam, spoon into hot, sterile jelly jars, seal, and process ten minutes in a boiling water bath.

Cel Hope, Sheridan, Wyoming

BUFFALOBERRY SPICY SAUCE

grated rind of 1 orange
2 cups of water,
2 cups of sugar
4 cups buffaloberries
¼ teaspoon ground cinnamon
pinch of ground cloves

Combine the grated rind of a fresh orange, 2 cups of water, and 2 cups of granulated sugar in a saucepan. Mix and cook over a moderate heat for 10 minutes. Add 4 cups of cleaned berries.

Cook until the berries pop. Now add ¼ teaspoon of ground cinnamon and a pinch of ground cloves and cook for 5 minutes. Stir frequently.

Spoon the mixture into a bowl and place in a refrigerator to chill. Serve chilled. This is a delightful red, spicy sauce and is best served with meat, as a flavoring.

Cel Hope, Sheridan, Wyoming

Black Currant

In my estimation, black currant is one of the most under-utilized wild berries in the West. It's a shame, because it is quite common and has a great flavor.

Black currant matures in mid-July through August, just when many families take fishing and camping trips. The nice part is that the currants are usually right next to streams that flow through the foothills and plains.

While many people prefer to fish on their vacations, it has been my experience that fishing isn't too good during the middle of the day and it's a pleasant break to do something different.

One time my sons, a friend, and I were camping along the Green River in southwest Wyoming. The fishing had been good, but after three days of float fishing in 90-degree heat, we were tuckered out. We elected to spend the day kicking around, so we walked a bit of the Oregon Trail, then hunted fossils, and finally, picked black currants.

The currants grew next to the Green, so they were big, plump berries. It took about an hour for the four of us to pick two gallons of currants. The currants kept well in my cooler and I was able to make them into jelly when I got back to Sheridan.

The currants rounded out a fine trip. We caught plenty of nice trout, found some good fossils, had an enjoyable trek on a historic trail, and had some luscious jelly for the winter. The varied trip appealed more to my 12-year-old sons than a strictly ''fishing-until-we-dropped'' type of vacation.

While you may prefer to fish until you drop, perhaps your spouse and children don't. I have heard a lot of fly fishermen that I have guided say ''I'd love to bring my wife and kids here, but there is nothing for them to do because they don't like to fish.''

Most western trout streams have a bounty of wild berries that grow along them. I would venture a guess that a day of berry picking, followed by a sight-seeing trip to nearby historical sites, mountain ranges, parks, and so on, followed by a trail ride on the third day would keep most non-anglers happy. I can't think of a famous river in the West that doesn't offer such nearby sidelights for non-anglers. In fact, Yellowstone Park streams offer even more sidelights, like naturalist talks, geysers, wildlife, and water-falls. The gist of this paragraph is that there is plenty to do besides fish, so bring the family along!

Black currant belongs to the genus *Ribes*. Other members of the genus include golden currant, squaw currant, sticky currant, gooseberry, and thin-flowered currant. All the members of the genus retain part of the flower at the tip of the fruit (see my explanation under gooseberry). The dried flower part looks like a little pigtail attached to a globe. (All the members of the genus have globe-shaped berries).

There are numerous species of *Ribes* in the mountain West. Dr. Dee Strickler states that there are fourteen species in Montana.

While some species of currant and gooseberry are more common and have more berries than other species, none are poisonous. (Squaw currant, *R. cereum*, is listed as nonedible. It has bright red berries which have little flavor. It occurs on sunny, dry slopes in the foothills and mountain regions). Some are more desirable than others, but none will kill you or make you ill. Essentially, if you see a globular berry with an attached pigtail, it's safe to eat.

Black currant likes deep, well-drained soils. You will find it growing along creeks and streams from the plains into the foothills. It also grows in protected draws and ravines in the plains and foothills. You can find it in every state in the mountain West.

Black currant blossoms in late April through mid-May. Its yellow, trumpet-shaped flowers produce a spicy fragrance that I find to be one of the most delightful natural fragrances I know of. It smells much like cloves.

The berries ripen in late July through August. They turn from green to red to a glossy black. (Some bushes will have golden-yellow fruit). The size of the berry ranges from pea-size to three-fourths the size of a marble. The berries can occur in clumps of up to ten, and many times there will be a range of green to ripe berries in the clump. The reddish and purplish currants will add more pucker power to the jelly you are making, so don't be afraid to pick them.

Black currant bushes range from three to six feet tall. The younger branches are slate gray in color, while the older branches are somewhat roughened and nearly black in color.

The leaves are simple, light green, and have a shape like a small maple leaf with two or four lobes.

Your chances of finding enough black currants to make a batch of jelly (say one gallon) are three out of five in normal years. In drought years, your chances diminish to one out of five. Be careful when you are picking black currants; I have found bushes that red ants have really taken a liking to. The ants will protect the currants from any intruder—humans included. Those little red devils can bite like the dickens. 🦌

Recipes

CURRANT JELLY

Select firm, not overripe fruit, wash it thoroughly, and remove the leaves but not the stems. Crush the fruit to start the juice and then heat it quickly without adding water. Cook and stir constantly from 5 to 8 minutes, or until the skins of the fruit are white. Strain through several thicknesses of cheesecloth. Do not squeeze the bag, but press lightly to start the flow of juice as it cools. Never make more than 4-6 cups of juice at one time.

To each cup of currant juice, add 1 cup of sugar. Stir until the sugar is dissolved, then bring quickly to boiling point in a 6- to 8-quart pot. Currants have so much pectin that, as a rule, just boiling up once will give the jelly test, that is, the juice will sheet from the spoon. Remove the pot from the heat as soon as the jelly test is reached. Let stand a minute or two, skim off the foam. Pour the jelly into hot sterilized glasses, cover with cheesecloth, and let stand until set. When cold, cover with melted paraffin and rotate so that a rim of the paraffin reaches the top of the glass. Cover with lids, label, and store in a cool place.

Laale Cina, Cody, Wyoming

CURRANT ICE CREAM SAUCE

1 cup washed and stemmed currants
½ cup sugar or ⅓ cup honey

Cook currants in water for 10 minutes. Add sugar or honey and boil gently for 5 more minutes. Serve hot or chilled over vanilla ice cream.

Margaret and Charles Butterfield,
Preserving Wyoming's Wild Berries and Fruits,
University of Wyoming Agricultural Extension Service, 1981.

CURRANT PUNCH

Sweeten hot currant juice to taste, stirring to dissolve sugar. Cool. Add club soda or ginger ale at serving time. (Other fruit juices may be combined with the currant for a flavorful punch.) For a special touch add a small scoop of ice cream at serving time.

Margaret and Charles Butterfield,
Preserving Wyoming's Wild Berries and Fruits,
University of Wyoming Agricultural Extension Service, 1981.

BEST EVER MINCEMEAT

6 cups finely chopped cooked venison
10 cups chopped apples
2 cups beef suet, ground
2 cups white sugar
2 cups vinegar
4 cups raisins
2 cups currants
1 quart apple cider
2 tablespoons each cinnamon, cloves, nutmeg
½ teaspoon pepper
1 tablespoon salt

Boil for 1 hour, stirring often to prevent scorching. Pour in sterilized quart jars and seal. Pressure can at 15 pounds pressure for 25 minutes.

Enid Purcell, Billings, Montana.
From the Billings Gazette Cookbook, *September 25, 1983.*

CURRANT CORDIAL*

8 cups currants
4 cups sugar
1 quart vodka

Place ingredients in a gallon glass jar. Stir. Cap with a loosely fitting lid. Stir every few days. Brew a couple of months (1-2), strain and bottle.

*Other berries (chokecherries, gooseberries) can be used.

Charlotte Heron, Missoula, Montana

SPICED CURRANT JELLY

2½ pounds currants (mashed and cooked till soft, strain out juice)
2 tablespoons stick cinnamon and ½ tablespoon whole cloves—place in
a small cheesecloth bag.

Boil spices in juice for 10 minutes, then remove the spice bag. For each
cup of juice add ¾ cup of sugar. Boil to jelly stage. Pour into hot, sterilized
jars. Process in hot water bath for 5 minutes. Makes six jars.

From Kerr Book.
Submitted by Charlotte Heron, Missoula, Montana.

CURRANT PIE*

Pastry for a 2 crust pie (see serviceberry recipes)
4 cups black currants

Cut off stems and blossom ends of the currants. Roll out a pie crust and
arrange it in a pie pan. Place berries in pie pan on top of crust. Mix 2
cups sugar, 2 tablespoons flour or cornstarch, and 1 teaspoon nutmeg.
Sprinkle over currants. Put on top crust or lattice crust.

Bake at 375 degrees for 45-50 minutes. Put a cookie sheet under the pie
pan, since the juices usually run over if a top crust is used.

For less intense flavor, or if you only have 2 cups currants, fill out with
2 cups sliced apples. Apples extend gooseberries and currants very well.

*This pie can be made with gooseberries.

Charlotte Heron, Missoula, Montana

CURRANT-RHUBARB JAM

1½ pounds frozen rhubarb (1 quart)
1 pound currants (1 quart)
1 package pectin
8½ cups sugar

Remove stems and tails from currants, combine with thawed, chopped rhubarb. Mash thoroughly in a kettle, add pectin, and stir until dissolved. Heat to boiling. Add sugar, stirring constantly. Bring to a full, rolling boil, continue stirring. Boil for 4 minutes. Remove from heat, skim off foam, pour into sterilized jars, seal and process in a boiling water bath for 10 minutes. Yields six 8-ounce jars.

Cel Hope, Sheridan, Wyoming

Chokecherry

Chokecherry is one of the most popular wild berries in the Rocky Mountain region. Its popularity might rest with its widespread distribution. Another reason is that it has a distinctive, tart taste that makes it a hit with so many cooks and gourmets.

Chokecherry has a variety of uses ranging from wine to jelly to syrup to daiquiris. The unique flavor of chokecherry lends itself to many recipes that use a juice as a start.

Wildlife uses chokecherry, too. Game birds, songbirds, raccoons, deer, coyotes, and bears relish the cherries, while sharp-tailed grouse depend on the buds for a winter food source. Elk, deer, and moose browse the branches during the fall and winter.

During my childhood, chokecherries weren't part of my berry-picking experience. They grew in Michigan, but my mother contended that they were too much trouble to pick and process. Besides that they were tart and puckery; if she wanted cherries, she'd pick the pie cherries off our trees.

When I moved to Wyoming in 1966, I soon realized that there weren't any cherry, apple, or peach orchards. If I wanted any fresh fruits or berries, I would have to find wild ones.

Chokecherries were one of the first that I learned to pick. It seemed so easy to pick lots of them because the long clusters of cherries make it easy to pick a handful with each reach. It doesn't take long to fill a berry bucket with a rate like that.

My sons, Clint and James, found out early that chokecherry jelly tasted great on bread or toast. I have some hilarious photos of them when they were two years old with chokecherry jelly from ear to ear and throat to eyebrow.

Because they loved jelly, it was only a matter of five years before they started helping me pick chokecherries. Both seemed quite content to pick a quart of chokecherries and call it quits. I picked the lion's share of the cherries (five gallons or so) until the boys reached eleven. I decided that they were old enough to pull their weight (or pick their weight) so I told them to pick the chokecherries that were next door while I was at work. The boys decided that they weren't going to do it, so I decided I wouldn't either. That winter when they clamored for jelly, I told them that they didn't pick any chokecherries so I didn't make any jelly.

The next summer I didn't say a darned thing about chokecherries. One August evening I came home from work to find a five-gallon bucket full of chokecherries sitting in the kitchen. James and Clint announced that they had picked them and when was I going to make jelly?

Chokecherries are a great berry to plan a family outing around. They usually ripen in August—a perfect picnic month. Chokecherries are a family affair. I have seen families spend a Sunday afternoon picking chokecherries. The families will break for a picnic lunch and then move on to pick another chokecherry clump. By the end of the day, the youngsters are happy, tired, and elated by their full berry buckets, while the parents are happy for the time spent together as a family as well as full buckets.

Chokecherry bushes usually occur in clumps. It is most common to find several bushes growing together or in close proximity. They occur on the plains and prairies, into the foothills and lower mountains. If there is a little water, you can even find them growing in the coulees and ravines of desert areas.

Chokecherries occur from all the northern Rocky Mountain states all the way east to the Atlantic coast. Chokecherry belongs to the rose family and has the scientific name of *Prunus virginiana*.

Favoring deep, well-drained soils, chokecherries make do on gravelly soil or on clay. They prefer mesic habitats, that is, not too wet, not too dry, but somewhere in the middle of the moisture range. Chokecherry reaches a height of four to twelve feet in the mountain West. The branches and trunk are dark gray, and the younger parts of the shrubs have a glossy copper tone. The branches have prominent lenticels (breathing pores), which look like small slits. The simple leaves are alternately arranged, bright green on top and dull green underneath, and taper from both ends to the middle. The leaf margin is serrate (small-toothed).

Depending on the altitude, chokecherries blossom anywhere from the first week of May to the first week of June. With the extended blooming period, you can enjoy the springtime floral display from the plains to the mountains. The numerous white blossoms occur in linear, drooping clusters called racemes. Often the blossoms will be so thick in a clump of chokecherries that the clump appears as a white curtain of flowers.

Chokecherries ripen sometime between early August to mid-September—again, it all depends on the elevation. This extended ripening period will allow you to pick chokecherries for more than a month. Chokecherries growing at lower elevation will ripen much sooner than those at seven thousand to eight thousand feet. When chokecherries ripen, the color turns from light green to wine red to glossy black. The ripe fruit ranges from pea-size to nearly marble-size.

Your chances of getting enough chokecherries to make jelly are excellent. In all years except extreme drought ones, your chances will be ninety-nine percent that you will be able to pick at least two gallons or more of chokecherries. In most years, you should rate eight chances in ten that you can pick five gallons or more. I won't rate your chances for getting ten gallons because rendering chokecherries into juice is so darned much work that I'd hate to think of anyone rendering more than ten gallons of chokecherries. Essentially, in a good year and in the right locale, you could pick all the chokecherries you could carry out.

One note of caution. Chokecherry leaves and shoots in the spring and summer are cyanogenetic. That is, under certain conditions they are capable of producing hydrocyanic acid. Chokecherry can kill livestock that consume the leaves and shoots. Some texts state that the hydrocyanic compound is present in the kernel of the cherry. The only part of chokecherry that you can eat safely is the fleshy outer part of the cherry. Since rendering chokecherries involves simmering them for fifteen minutes or longer, the hydrocyanic compound in the kernel becomes harmless—but toss the pits out after you render the cherries anyway, just to be on the safe side.

I find that I can save chokecherries for later use by either freezing them whole in plastic freezer bags, or by rendering the juice and freezing the juice in milk cartons or jugs.

Recipes

Probably the most difficult part of making any chokecherry recipe is obtaining the juice from the cherries. I usually place a layer of chokecherries in an 8-quart or larger kettle, mash them, add another layer, mash them, and so on, until the kettle is one-fourth to one-third full. I add a cup or two of water, cover, and bring it to a boil, then I simmer it for fifteen minutes or so. Depending on what equipment I have on hand, I either let the cooked chokecherries cool, or start pressing them to extract the juice.

A Foley food mill comes in handy to extract the juice. Other folks use a jelly bag and let the juice drip out and then squeeze the bag after an hour or so to get the remaining juice. Sometimes I place a colander over a large bowl and add a cup or so of the cooked cherries. Using a heavy metal spoon, I can press a lot of the juice. Nancy Brannon contends that the easiest way of extracting chokecherry juice is with a steam extractor. She states that she uses over 150 pounds of chokecherries a year and the steam extractor makes the chore a snap. Whatever method you use, realize that extracting the juice takes time, so don't start this project until you have plenty of time.

CHOKECHERRY SYRUP

Clean stems and leaves from berries. Place in large kettle. Cover with water and cook. When berries are pulpy, remove and pour juice and berries into press.

Boil equal amounts of juice and sugar. Boil for ten minutes or so. Do not boil too long, since it will jell. Pour into clean half-pint jars and seal. Process in a boiling water bath for ten minutes to ensure the syrup is preserved.

Kay Ellerhoff,
Montana Outdoors, "Cooking the Wild Berry,"
July/August 1975

CHOKECHERRY SYRUP

4 cups chokecherry juice
4 cups white sugar
2 cups white corn syrup

Place all ingredients in a 5-quart kettle. Bring to a boil, stirring to dissolve sugar. Turn heat to medium and continue boiling for 10 to 15 minutes or until foam starts to climb sides of kettle. Remove from heat and pour into sterilized jars and seal. After opening, it will keep in refrigerator for weeks.

Marjorie Helms and Alice Halsted,
Sheridan, Wyoming

PIONEER CHOKECHERRY SYRUP

4 cups chokecherry juice
4 cups sugar
1 teaspoon cream of tartar

Cook over medium heat until mixture coats the spoon (like gravy does). Refrigerate for immediate use or pour into clean hot jars and process in boiling water bath, 10 minutes for pints.

Margaret and Charles Butterfield,
Preserving Wyoming's Wild Berries and Fruits,
University of Wyoming Agricultural Extension Service, 1981.

CHOKECHERRY LIQUEUR

1 quart gin
1 quart washed chokecherries
3 scant cups sugar

Put all ingredients in large container (say a 2-quart jar).

Shake every day for 30 days. Strain though cheesecloth or a diaper. Bottle. Batch can be doubled using a gallon jar.

Pat Stevens, Cowley, Wyoming

WILD CHOKECHERRY DAIQUIRI

For the chokecherry syrup:
　　4 cups fresh chokecherry juice
　　4 cups sugar

For the daiquiris:
　　½ cup chokecherry syrup
　　¾ cup sweet-and-sour bar
　　　　concentrate
　　1 cup rum
　　About 3 cups ice cubes
　　　　or crushed ice

Combine the chokecherry juice and the sugar in a large saucepan and bring it to a rolling boil over high heat. (At this point, you may want to can the syrup, using traditional canning procedures. In that way, it will keep for a year or more on your pantry shelf.) Chill the syrup in the refrigerator prior to mixing the daiquiris, or it will not make the "slush" type drink this is supposed to be. Then, combine all the ingredients for the daiquiri in a blender. Blend on high speed for about 30 seconds, or until the ice is all incorporated into the cocktail. Pour into suitable glasses, and garnish with either a spiral of fresh lime, or a sprig of fresh mint. Serve.

Storage: Store the blended daiquiris in a covered plastic container in the freezer for up to a week.

Preparation time: Syrup, 10 minutes. Daiquiris, 10 minutes.

Nancy and Dave Brannon, Feasting in the Forest, *1989.*

CHOKECHERRY PIE

1 (9-inch) baked pie shell
2 cups chokecherry juice
3 level tablespoons cornstarch
1 cup sugar
small pinch of salt
½ teaspoon almond extract

Cook until thick, stirring constantly. Cool. Pour into pie shell, chill. Serve with whipped cream or cream topping.

Bob Giurgevich, Sheridan, Wyoming

CHOKECHERRY/CRAB APPLE JELLY

chokecherries
crab apple juice
sugar

Put 2 quarts chokecherries in a deep pan. Cover with ½ inch water. Boil 3 minutes or until they break open. Strain through a wire mesh.

To 4 cups chokecherry juice, add 1 cup crab apple juice and 1 box MCP pectin. Bring to boil and add 6 cups sugar. Boil 3 minutes until sugar dissolves. Pour into jelly glasses and seal.

Lynda Nelson, Columbia Falls, Montana

CHOKECHERRY/CRAB APPLE SYRUP

4 cups chokecherry juice made as above
2 cups crab apple juice
6 cups sugar
1 envelope Certo

Bring to a boil to dissolve sugar. Put in hot jars and seal.

Lynda Nelson, Columbia Falls, Montana

CHOKECHERRY/APPLE BUTTER

4 cups apple pulp
2 cups chokecherry pulp
5 cups sugar
½ teaspoon almond extract

Prepare pulp of both fruits first by putting cooked fruit (unsweetened) through a sieve or food mill. Heat to a boil, stirring carefully. Add sugar. Stir constantly until it just begins to thicken. Add extract and blend. Ladle into hot jars. Adjust lids at once and process in boiling water bath (212 degrees F.) 5 minutes. Remove from canner and complete seals unless closures are self-sealing type. Makes 8 half-pints.

Bob Giurgevich, Sheridan, Wyoming

OLD TIME CHOKECHERRY JAM

Remove stems from chokecherries and wash. Drain. Add 1 cup water to every 4 cups fruit. Place over low heat and simmer until fruit is very tender, stirring occasionally. Rub pulp through a medium sieve; measure, and add an equal amount of sugar. Place over moderate heat and stir until sugar has melted. Bring to a full, rolling boil and cook until mixture sheets (220 F.). Stir constantly. Seal in hot sterilized jars. Three cups pulp will make about 3 half-pints.

Bob Giurgevich, Sheridan, Wyoming

CHOKECHERRY SAUCE

1 quart gin
2 cups unpitted chokecherries
1 cup sugar

Put in a half-gallon glass jar, then let set till Christmas (about 4 months). Great for milkshakes or over a dish of ice cream.

Dean Davis, Sheridan, Wyoming

BEFORE BREAKFAST DRINK

Put 1 cup raw chokecherries into a quart jar that has been sterilized. Add 1½ cups sugar and fill to within 1 inch of the top with hot water. Seal and process in hot water bath for 20 minutes.

A glassful before breakfast is an excellent remedy for colds or as a tonic.

Mrs. Frank Ricki, Lewistown, Montana.
From the Billings Gazette Cookbook, *September 25, 1983.*

CHOKECHERRY BOUNCE

Fill 1 gallon glass jar three-fourths full of chokecherries; add 3 cups sugar and fill jar to top with vodka (about 2 fifths).

Let stand at room temperature about 4 months. Strain off juice through cloth. Do not squeeze or you will have pulp in it.

Makes about 3 fifths and 1 pint. Do not use metal caps on bottles; use corks or plastic wrap and rubber bands.

This is similar to sloe gin and makes a good mixed drink with a lemon-lime bottled drink.

Mrs. Erna Geving, Baker, Montana.
From the Billings Gazette Cookbook, *September 25, 1983.*

CHOKECHERRY WINE

Wash and grind fruit (through coarse grinder) and add boiling water—1½ parts water to 1 part fruit. Measure fruit before grinding.

Let stand overnight and strain off juice and put into large crock. Add 3 pounds sugar to 1 gallon juice and dry yeast (1 package to 5 gallons liquid). Mix well and cover crock with plastic wrap. Put one end of plastic tube into liquid and the other in jar of water. Let stand in warm place until juice has stopped working. Siphon into bottles and let stand for 1 year.

If sweeter wine is desired, add more sugar.

Mrs. H. J. Klindt, Billings, Montana.
From the Billings Gazette Cookbook, *September 25, 1983.*

CHOKECHERRY ADE

3 to 4 tablespoons chokecherry syrup
1 to 2 tablespoons lemon juice
Ice and water to fill glass

Variations of the drink include adding crushed mint leaves or omitting lemon juice and using ginger ale in place of some of the water.

Ruth Scharff, Bozeman, Montana.
From the Billings Gazette Cookbook, *September 25, 1983.*

HOT SPICED CHOKECHERRY DRINK

Per cup:
    2 tablespoons chokecherry syrup
    1 tablespoon lemon juice
    ½ to ¾ cup water

Heat with a bag of whole spices (cloves and cinnamon). Remove spices when flavored enough.

Variation: Omit lemon juice and substitute cider or bottled apple juice for it and part of the water. This drink is also very good chilled.

> *Ruth Scharff, Bozeman, Montana.*
> *From the* Billings Gazette Cookbook, *September 25, 1983.*

CHOKECHERRY PUDDING

Bring pot of water to boil, put several cups of fresh or frozen chokecherries into pot and boil for about one hour, to get the juice out. Strain off the juice and throw away the pulp and pits. Mix ¾ cup flour with 1 cup of water, stir into a paste, gradually add this to the chokecherry juice, keep stirring while the mixture simmers on the stove. As it thickens, add some sugar to sweeten. Before taking the saucepan off the stove, add two tablespoons butter to keep pudding from coagulating. Serve with Indian fry bread. Served with deer jerky, this makes a typical dinner for the Crow Indians.

> *Theo Hugs, Ft. Smith, Montana*

CHOKECHERRY-APPLE JAM

1 quart chokecherry pulp (approximately 4 quarts of chokecherries)
1 quart unsweetened applesauce
9 cups sugar
¼ cup lemon juice
1 package pectin
¼ teaspoon margarine

Bring to a boil 4 quarts of washed chokecherries with 1 cup of water, cover and simmer for 20 minutes. Run the cooked chokecherries through a Foley food mill or rub through a colander.

To a 6- or 8-quart saucepan or kettle, add 1 quart chokecherry pulp, 1 quart unsweetened applesauce, 9 cups sugar, ¼ cup lemon juice, and 1 package pectin. Mix thoroughly to make sure the pectin and sugar are dissolved. Heat to boiling, add margarine to prevent foaming. Boil for four minutes, skim, pour into sterilized jars, seal and process in a boiling water bath for 10 minutes. Yield: ten 8-ounce jars

Cel Hope, Sheridan, Wyoming

Elderberry

Elderberry brings back many childhood memories. Elderberry thrived in the rich mucklands of south-central Michigan where I grew up. Elderberry matured in the fall about the time that we dug the potatoes and carrots and stored them in the root cellar. It was time to make sauerkraut, pick up the windfallen apples, and make cider.

We picked elderberries just before the hickory nuts started to fall. It seemed that we would go out on the nicest Sunday afternoon imaginable, with temperatures in the mid-60 degree range, no wind, and plenty of bright sunshine. We would find the elderberry bushes quite easily, for they seemed to tower over the rest of the vegetation in the mucklands. Each of us would have a paper grocery sack which we would fill with the clumps of elderberries—we would pick the whole clump and wait until we got home to pull the individual berries off the stems.

Mom got busy with the elderberries and made them into pies and jelly. That was the nice thing about elderberry picking—the rewards (pies and jelly) were quick and great-tasting.

When I moved to Wyoming in 1966 to attend the University of Wyoming, I thought my berry-picking days had come to an end. I didn't see how such a barren, dusty, windblown place could have any berries. It took a couple of years, but I found that Wyoming and the other western states had plenty of berries. When I moved to Logan, Utah, in 1969 to continue my graduate work, I found that there were berries growing there that grew in Michigan. Elderberry grew in profusion in a canyon only a few miles away from the Utah State campus. That year was one of the first that my former wife and I made jelly. I found enough elderberries for her to make a decent batch of jelly, which we gave as Christmas gifts that year.

There are two species of edible elderberries in the Rocky Mountain region: blue elderberry and black elderberry. A third species of elderberry, red elderberry, is poisonous. All three species belong to the genus *Sambucus*. Blue elderberry is *S. coerula*; black, *S. melanocarpa*; and red, *S. microbotrys*.

Elderberries like moist, well-drained, deep soils. That limits them to floodplains and moist slopes, from the foothills to the subalpine regions.

Elderberry bushes grow in clumps where there might be ten to thirty canes (stems) growing in the clump. The cane is quite pithy and there are very prominent lenticels (breathing pores) on the bark. Elderberry shrubs range from four to twelve feet high. Elderberries have pinnately compound leaves, which have from five to eleven leaflets. The color of the leaves tends toward medium to dark green. The leaflets are glabrous (that is, they lack hairs so they have a smooth, glossy appearance). The leaflets are serrate (toothed) and are lanceolate (lance-shaped).

Elderberry blossoms in June. The blossoms are white and are clustered together in a flat or slightly convex arrangement (called a "cyme"), ranging from silver-dollar size to saucer-size. Black elderberry has a slightly convex flower arrangement; blue elderberry is flat.

Elderberry ripens in September. Black elderberries have a glossy black color and range in size from BB-size to split pea-size. Blue elderberry has a powdery blush—it is actually black but the blush gives it a powder blue look. The berry is the same size as black elderberry.

Incidentally, red elderberry has orange or red berries and usually ripens in August and grows in the mountains. It has high concentrations of saponin, which causes animals to avoid it. If you remember that edible elderberries don't ripen until September and they are either powder blue or black, then you won't have any troubles with red elderberry.

Eating raw blue or black elderberries can cause nausea, but cooking the berries eliminates the problem. In other words, don't eat raw elderberries!

Your chances for finding enough elderberries to make a pie (two to three cups) are good (ninety percent) if you live in western Montana, eastern Idaho, or northern Utah. In other places, your chances are hit or miss—if you find a good elderberry clump, you will probably get enough for a pie. However, you probably have a one in twenty chance of finding such a clump. Finding enough berries for jelly (four quarts) is again good if you live in the above areas. If the birds don't beat you to it, I would say that your chances are four out of five.

Remember to pick the entire fruit clump, go home, sit down, and pull off the berries. This is a much easier process than trying to pick the single berries in the field.

You could freeze elderberries, but usually a picker will only get enough for a pie or jelly. I would be more inclined to freeze the juice than the berries. That way I could make jelly or wine when I got around to it. 🍒

Recipes

ELDERBERRY JELLY

4 quarts fresh elderberries to yield 4 cups juice
6 cups water
6 cups sugar

Pick over berries, removing all stems and leaves. Do not worry if you have a lot of underripe berries because these have more pectin. Wash berries under running water. Put in 8-10 quart kettle. Add the 6 cups water and bring to boil over high heat. Reduce heat to moderate and cook uncovered for 45 minutes, stirring from time to time. Put a large sieve over a large pot, line sieve with double cheesecloth and pour in the berries. Allow the juice to drain into the pot without disturbing.

When the juice has drained through completely, measure it (you need 4 cups) and return to the pot. Discard the berries. Add 1½ cups sugar for each cup of juice and bring to boil over high heat, stirring until the sugar dissolves. Boil uncovered until the jelly reaches 220 F. on a candy thermometer. Remove from heat, skim off the surface foam. Pour into hot, clean jelly glasses, and seal with paraffin. (The USDA recommends the boiling water bath for preserving jams and jellies).

Bob Giurgevich, Sheridan, Wyoming

ELDERBERRY AND APPLE JELLY

1 cup diced apples (cored and peeled)
1 cup elderberries
2 cups sugar

Wash and drain fruit. Add enough water to keep from burning. Cook until apples and elderberries are tender.

Put through sieve and drip through jelly bag. Add sugar and cook until jelly stage is reached. Put into glasses and seal with paraffin.

Bob Giurgevich, Sheridan, Wyoming

ELDERBERRY AND GRAPE JELLY

elderberries
sugar
bottled unsweetened grape juice

Wash and stem the elderberries and pick them over. Put into deep pan, add enough water so that you can see it through top layer, and bring to a boil. Cook until berries are soft. Strain through jelly bag. Measure juice into deep pan, then measure out and add an equal amount of unsweetened grape juice. For each cup of the mixed juices add ¾ cup of sugar. Bring to a boil, test frequently for jellying. When 2 drops run together off side of spoon, pour into glasses and seal with paraffin.

Bob Giurgevich, Sheridan, Wyoming

ELDERBERRY SYRUP

Wash berries and cover with water. Boil, mashing berries as they boil for 8 to 10 minutes. Strain. Put 4 cups juice in pan. Bring to boil. Add 4 cups sugar. Boil until thickness you prefer for syrup. Will thicken a little as it cools.

Gladys Christensen, Cooke City, Montana.
From the Billings Gazette Cookbook, *September 25, 1983.*

WILD FRUIT DUMPLINGS*

Stew mixture:
    2 cups elderberries
    1½ cups water
    ¾ cup sugar

Combine berries, water, and sugar. Stew until well-cooked.

Dumplings batter:
    ½ cup flour
    2 teaspoons baking powder
    dash salt
    1 tablespoon sugar
    1 egg
    ¼ cup milk

Mix well. It should be fairly thick. Drop by teaspoonsful into boiling elderberries; cover tightly and simmer 15 minutes. Serve with a hard sauce made by creaming together 1 cup brown sugar and ¼ cup butter or margarine. This recipe makes about 4 medium dumplings; it may be doubled. Best served when hot.

*Huckleberries, low bush blueberries, serviceberries, or wild strawberries can also be used.

Fran Davis, Otis Orchards, Washington. From Savoring the Wild, *A collection of favorite recipes from the employees of the Montana Department of Fish, Wildlife, and Parks. Falcon Press, Helena, Montana, 1989.*

ELDERBERRY WINE

Mash 20 pounds of berries in a 5-gallon crock. Add 5 quarts boiling water. Cover and let stand 3 days. Strain juice and return it to crock. Add 10 cups sugar. Let stand until fermentation ceases. Remove scum. Strain and bottle. Let age for 1 year.

Charlotte Heron, Missoula, Montana

Gooseberry

Gooseberry is widely distributed in the mountain West, but few folks recognize the berries or realize that they are very edible. It seems that a person should be able to recognize this bush quite easily because the branches are loaded with ¼-inch to ½-inch thorns.

During late July, 1989, I was guiding a Colorado couple on the Big Horn River. Bill and Pam had fished with me for several years and we had become friends. They were fun to guide because they didn't have to catch hundreds of trout to be happy. They enjoyed viewing the wildlife and plant life along the river almost as much as fishing.

On this particular day, Bill needed to make a rest stop on an island. I continued to help Pam fish a pool about thirty yards away from the island. About ten minutes later Bill waded to us. He opened his hand and said "What are these?"

I looked at the berries in his hand: half were Solomon's seal, the other, gooseberries. I swept the former into the river, then I told Bill, "These are gooseberries. They are great for jam and pies."

Bill responded that the island was covered with them. It was lunchtime, so I suggested that we eat on the island. After lunch, I found that there were a dozen or so gooseberry bushes within twenty yards of where we sat. I picked a few to eat and thought that was that. To my surprise, Bill said, "Can we pick some gooseberries? We've never done any berry picking and we would like to try it."

I said "sure." I waded back to my drift boat, got my gallon milk container berry bucket, and we started to pick. In a half hour or so, we managed to pick two-thirds of a gallon of gooseberries. Bill and Pam raved about how much fun it was. They thought that picking the berries added to the overall enjoyment of their fishing trip.

I guess that is the nice thing about berry picking—it doesn't have to be the only reason to get out and enjoy the out-of-doors. Berry picking is a nice sidelight to about any outdoor recreational activity in the summer and fall—an added bonus to your fishing trip, hiking trip, birding expedition, hunting trek, or picnic. Berry picking can be something you do when the fishing is slow during midday. I know that getting a bucket of berries is almost as satisfying for me as catching a bragging-sized trout.

By the by, Bill and Pam came back two weeks later. I had baked a gooseberry pie for them which they devoured. They now are hooked on berry picking and have resolved to pick berries the next time they fish with me. They are trying to schedule this summer's fishing trip when the chokecherries ripen.

Gooseberries are quite common in the Rocky Mountain region. There are several species here, but the one I encounter most frequently goes by the common name of white-stemmed gooseberry. Its scientific name is *Ribes inerme*. The specific name, *inerme*, is really a misnomer for it means "unarmed" and gooseberry is definitely armed with prickles and spines. In fact, the spines are how I recognize gooseberry bushes during the winter and early spring.

Gooseberries grow in floodplains along stream courses throughout the mountain West. I have found them in Jackson Hole in the Gros Ventre slide area, along Fish Creek and parts of the Snake River floodplain. Some of the canyons in the Big Horn Mountains have high concentrations of gooseberries. The same holds true for most mountain ranges in the West.

Gooseberry blossoms in May. It has a white to pinkish trumpet-shaped flower. Gooseberries ripen mid-July to mid-August. They turn from green to reddish-tinged to deep purple color. The green berries have white lines in them; as the wild berry ripens the lines become less distinct. (The lines remain quite distinct in domestic gooseberries).

One identifying feature of gooseberries and currants is that they retain part of the flower. This dried flower part looks like a pigtail at the point on the berry opposite its stem. I used to tell folks that flower parts were attached at the bottom of the berry until a botanist from Maine, Sam Ristich, pointed out that the flower parts are actually at the top of the berry. He explained further that the gooseberry flower has an inferior ovary, that is, it is located beneath the floral parts. When the ovary is fertilized it expands and pushes the dried flower parts to the top of the berry.

Gooseberry leaves are three- or five-lobed and look a lot like rounded maple leaves. The leaves are less than 1½ inches broad.

Gooseberries and currants grow throughout the mountain West, but that's not the case in the Northeast. Gooseberries and currants are intermediate hosts for white pine blister rust. Consequently, foresters have tried to eradicate all currants and gooseberries to save the forests. The effort was a total failure.

Gooseberries make great pies, jams, and jellies. In order to make a pie you will need about a quart of berries, and a little over two quarts to make jam or jelly. Most of the time you can pick two quarts of gooseberries in two hours' time. In a normal year, your chances of getting two quarts of gooseberries are two out of three.

Gooseberries store easily. Wash them in cold water, let them drain dry in a colander, then place them in quart plastic freezer bags and put them in the freezer. They will keep well for six months or so. 🍒

Recipes

GOOSEBERRY JAM

4 pounds gooseberries or currants
4 pounds sugar
¼ teaspoon salt

Remove the stems, or the "tops and tails" from gooseberries. Wash the berries, crush and cook until fairly tender. Add the sugar and salt and continue cooking until thick. If the berries are ripe and not too acid, three-fourths as much sugar as fruit can be used (3 pounds sugar). Pour into hot sterilized glasses or jars and seal, label, and store in cool place.

Laale Cina, Cody, Wyoming

GOOSEBERRY JELLY

Wash fruit well. Remove any remaining blossom ends. Place fruit in pan and add ¼ cup of water for each cup of berries. It may help to release juice by lightly crushing the berries. Extract juice by boiling water/fruit mixture approximately 10 minutes. Pour cooked fruit into jelly bag and allow to drain. Press bag to get all of the juice.

Measure juice into large pan. Add 1 cup of sugar for each cup of juice. Heat quickly to boiling, stirring to dissolve sugar. Continue boiling to jelly stage. To test, dip large spoon into boiling syrup; lift spoon so syrup runs off side. When syrup runs off spoon in two distinct lines of drops which "sheet" together, stop cooking. Pour into scalded jelly jars, adjust covers and process* in hot water bath for ten minutes.

*As of 1988, USDA recommends that jams and jelly should be put in regulation jars, sealed, and processed in hot water bath 10 minutes.

Mrs. Pauline Deem, Plentywood, Montana

GOOSEBERRY PIE

⅔ cup water
2 cups sugar
1½ quarts fresh gooseberries (4 cups if berries are large)
¼ cup cornstarch
Pastry for 2-crust pie (see serviceberry recipes)

Cook ⅓ cup of water and sugar in saucepan over low heat 2 or 3 minutes. Add berries. Simmer gently about 5 minutes until cooked but still whole. Using a strainer, drain syrup from berries into small pan and place berries in pie shell.

Dissolve cornstarch in remaining ⅓ cup of water. Stir into the syrup. Cook over moderate heat until thick and clear, stirring constantly, about 3 minutes. Cool to lukewarm and pour over berries. Cover with top crust. Bake in 450 degree oven for 10 minutes then reduce heat to 350 for 35 minutes or until golden brown.

Mrs. Pauline Deem, Plentywood, Montana

CANNED GOOSEBERRIES

Canned gooseberries can be eaten as sauce or used in pies. Wash and stem berries. Add ½ cup water for each quart of fruit. Cover pan and bring to a boil. Stir occasionally to prevent sticking.

Pack hot berries to ½ inch from top. Adjust jar lids. Process in boiling water bath 10 minutes for pints, 15 minutes for quarts. Add sugar before making into pie or when serving as a sauce.

Margaret and Charles Butterfield,
Preserving Wyoming's Wild Berries and Fruits,
University of Wyoming Agricultural Extension Service, 1981.

GOOSEBERRY SYRUP

3 cups ground ripe gooseberries
½ cup water

Place in a saucepan and simmer on stove for 10 minutes. Extract juice.

2½ cups juice
2½ cups sugar
1 cup white corn syrup
½ pouch liquid pectin

Place in large saucepan and bring to a boil. Use a candy thermometer and boil to 210 degrees. It takes about 2 minutes. Skim foam off syrup with a large metal spoon. Put into sterilized jars and seal. Makes 4 (half-pint) jars.

"A beautiful clear red syrup that is excellent on pancakes or ice cream."
Janice Larson, Billings, Montana.
From the Billings Gazette Cookbook, *September 25, 1983.*

WILD GOOSEBERRY DUMPLINGS (SLUMP)*

Sauce:
 2 cups wild gooseberries
 1½ cups water
 ¾ cup sugar

Dumplings:
 1 cup biscuit mix
 2 tablespoons sugar
 ½ teaspoon nutmeg
 ⅓ cup milk

In a 2- or 3-quart saucepan, mix gooseberries, water and ¾ cup sugar. Bring to boil. Cover, reduce heat, and simmer for 10 minutes.

Mix dry ingredients for dumplings. Add milk and mix well with a fork. Drop batter by tablespoonsful atop fruit mixture. Cook uncovered over low heat 10 minutes; cover and cook 10 minutes longer. Makes 4 to 6 servings.

This dessert can be made while camping if you have a saucepan with lid.

* Huckleberries, low bush blueberries, and wild strawberries can be used for this recipe.

Wilora Dolezal, Cody, Wyoming.
From the Billings Gazette Cookbook, *September 25, 1983.*

GOOSEBERRY CONSERVE

1½ quarts gooseberries (stem and blossom ends removed)
1 cup raisins
¾ cup seeded and chopped orange (about 1 medium)
4 cups sugar

Combine all ingredients in a large sauce pot. Bring slowly to a boil, stirring until sugar dissolves. Cook almost to jellying point, about 30 minutes. As mixture thickens, stir frequently to prevent sticking. Pour hot into jars, leaving 1/4-inch head space. Adjust caps. Process 15 minutes in boiling water bath. Yield: about 6 half-pints.

From Ball Blue Book, *1990 edition.*
Submitted by Charlotte Heron, Missoula, Montana.

SPICED GOOSEBERRIES

2 cups cider vinegar
2½ pounds brown sugar
1 tablespoon each ground allspice, ground cloves, ground nutmeg, and
   ground cinnamon
2 teaspoons salt
4 pounds gooseberries (stems and blossom ends removed)

Put all ingredients except berries into a large pot and heat to dissolve sugar. Boil 5 minutes. Add berries and simmer 30 minutes or until tender. Pour into hot, sterilized jars and seal. Process 10 minutes in hot water bath.

Charlotte Heron, Missoula, Montana

# Hawthorn

Sometimes I can be blind to the obvious. I don't know how many times I have seen thickets of hawthorn that were laden with orange, red, or black apples (haws) and not thought a darned thing of it. One day, a good friend from Jackson, Wyoming, Kathy Buchner, asked if I had ever had hawthorn jelly. I said no, that I didn't think they were edible. She informed me how wrong I was! She and her husband, Jay, had been picking them for years in Jackson Hole.

I had really missed out, for haws are so easy to pick. It is a cinch to get lots of haws because they grow right at eye level—eliminating stooping or reaching very far.

When I discussed black currants, I mentioned combining fishing trips with berry picking. It is just as easy to combine a berry-picking trip with hunting. The hawthorn-filled draws that are so common in northeastern Wyoming and eastern Montana are very attractive to sharp-tailed grouse and pheasants.

This fall I hunted one of my favorite places in southern Montana—a beautiful area with natural grasslands, sandstone hills topped with ponderosa pine, and draws thick with hawthorn, wild plum, chokecherry, and serviceberry. It was early October, and frost had shrivelled the wild plums. The chokecherries and serviceberries were past. The hawthorn, however, hung in clumps on every branch of every bush.

The sharp-tailed grouse were gorging themselves on the haws and my black labs, Midi and Millie, had an easy time of flushing enough birds for me to get my limit of three. As I walked back to the car, I realized that I had been too successful, it was early afternoon and I was done hunting. Then I realized that a lot of good jelly material was hanging on those bushes.

I grabbed my berry buckets from my vehicle, walked twenty feet to the nearest hawthorn bush and started picking. Within an hour and a half I had picked two gallons of the orange haws and was heading home. That was some afternoon—I got two limits: one of sharp-tailed grouse, another of hawthorn. I was doubly blessed that day for the same creator that gave me birds to sustain me also gave me fruit. Besides the food, the country gave me a feast for my eyes. The autumn colors had arrived and every draw, hillside, and valley had vibrant fall colors of shimmering red, riotous orange, phosphorescent yellow, and eternal green. I had limited out on game, berries, and beauty—a heavenly afternoon in my estimation.

Hawthorn is a common shrub throughout the mountain West. Some species grow along stream courses, some grow on hillsides, others like the draws and canyons in the plains. Most species like deep, well-drained soils. Some hawthorns have orange fruits (haws), others have red, still others have dark red or black haws.

One telltale feature of hawthorn is that it has long, unbranched thorns. The thorns are two to three inches in length. The thorns are the key identifying feature of the genus.

Hawthorn may grow anywhere from six to twenty feet in height. The leaves are variable depending on the species. All are simple leaves, most have deep serrations and some have lobes—it just depends on the species. The leaf color is medium green.

Hawthorn belongs to the genus *Crataegus*. There are at least four species growing in the mountain West: *C. succulenta*, *C. erythropoda*, *C. rivularis*, and *C. douglassi*.

Hawthorn blossoms May into June depending on the elevation. The blossoms are white, with ten or so per cluster. Each blossom has five petals and lots of stamens (the flower parts that contain the pollen).

Hawthorn ripens in late August into September. The haws look like miniature apples (hence, two common names, crab apple and thornapple). They range from pea-sized to half the size of a normal grape. At the tip of the fruit there will be a star-shaped growth of sepals. If you were to look closely at a regular apple, you would see the same growth. Haws usually occur in clusters of five to twelve. The haw stems are relatively long—two to three inches.

Hawthorn seems to bear decent crops year after year. If you find a thicket that doesn't have any haws, go to the next thicket. If you drive or walk a bit and find several hawthorn thickets, you will probably find one laden with haws. So, I would rate your chances of getting two gallons or more of haws at ninety-five out of one hundred.

Haws don't keep too well, so it would probably be better to render the juice and freeze it, if you don't have time to put it up immediately. I have found that milk cartons or jugs make great freezer containers. 🌰

Recipes

FLOWERING HAWTHORN APPLE JELLY

Wash the fruit thoroughly. Barely cover with water; boil until the fruit gives off its juice. Drain through a jelly bag. (You will need 5 cups of juice.)

5 cups hawthorn juice
7 cups sugar
juice of 1 lemon
5 drops oil of cloves
1 bottle Certo

Combine the juice and sugar. Bring to a full rolling boil. Add the lemon juice and oil of cloves. Add the Certo and boil 1 minute. Pour into jelly glasses and seal. This makes a delightful pale pink jelly that is delicious with meat.

The flowering hawthorn produces a bright rose-red fruit about the size of a pie cherry. Do not gather overly ripe fruit, as it will not jell properly.

Fran Davis, Otis Orchards, Washington.
From Savoring the Wild, *A collection of favorite recipes from the employees of the Montana Department of Fish, Wildlife, and Parks.*
Falcon Press, Helena, Montana, 1989.

HAWTHORN JELLY

1 pound hawthorn
½ cup lemon juice

Wash and crush ripe fruit. Add 1 cup water and simmer 15 minutes. Squeeze out juice. Measure 3 cups juice and ½ cup lemon juice.

1 package MCP pectin   4½ cups sugar

Mix 3 cups hawthorn juice and ½ cup lemon juice in a kettle with pectin and stir well. Place over high heat; bring to a boil, stirring constantly to avoid scorching. Add sugar; mix well. Continue stirring and bring to a full rolling boil (a boil that cannot be stirred down). Boil hard exactly 2 minutes. Remove from heat. Skim foam and pour into jars. Fill to ⅛ inch from top. Place lids on jars; tighten bands and process in hot water bath for 5-10 minutes.

MCP Pectin, Anaheim, California.
Submitted by Charlotte Heron, Missoula, Montana.
"I use the MCP recipe for chokecherry. This is my favorite jelly."

THORNAPPLE JAM

5 quarts thornapples, reduced to 7 cups pulp
¼ cup lemon juice
1 cup apple juice
½ cup water
1 package Slim Set pectin
1⅓ cups honey

Cook thornapples in 2 cups water until they begin to pop (about 20 minutes). Press the cooked apples through a sieve or food mill.

To a large saucepan or kettle, add pulp, lemon juice, apple juice, and water. Slowly add pectin. Stir until dissolved. Add honey. Bring to rolling boil. Boil 1 minute. Remove from heat. Pour into sterilized jars, seal, and process 10 minutes in boiling water bath. Yield: eight 8-ounce jars

Kathy Buchner, Jackson, Wyoming

Huckleberry

Huckleberry. The mere mention of it brings up images of all-American things. While not as nationally revered as apple pie, motherhood, and the flag, huckleberries are still pretty much part and parcel of the United States. Huckleberry Finn surely brought that point home.

In the West, huckleberries are prized above all other wild berries. Towns have festivals celebrating huckleberries. I'm sure you have heard of the Huckleberry Festival at Trout Creek, Montana.

The festival lures people in to try various huckleberry taste treats like muffins, pies, and cobblers. There are huckleberry pancake breakfasts at the Fireman's Hall, a Little Miss Huckleberry Pageant, an arts and crafts festival, old-time fiddlers, horseshoe contests, and chuckleberry laugh-offs.

All the hoopla is due to the huckleberry patches around Trout Creek. They are so good that Trout Creek was declared "Huckleberry Capital of Montana" by the 1981 Montana State Legislature.

My first recollections of picking western huckleberries was when we resided in Jackson Hole in 1971. My former wife, Peggy, and I were renting a cabin from a Mormon couple who homesteaded on Antelope Flats in 1912. John and Barthie Moulton had accumulated a wealth of historical and practical knowledge that included every berry patch in Jackson Hole. While they knew the spots, they weren't apt to share the information with just anyone, particularly a pair of outsiders.

Fortunately, our sons, James and Clint, were but a couple of months old when we moved in. Barthie, being a big-hearted grandmother many times over, took Peggy and the boys in as her own. She would introduce the boys as "her grandsons." Barthie wanted to make sure that the boys were well taken care of, so she would help Peggy out. When there were fresh berries to be had, Mrs. Moulton would make sure that we had some. The berries usually came in the form of pies or cobblers that she made in the oven of her Monarch wood cookstove.

Perhaps the biggest honor came when the Moultons elected to share with us their favorite huckleberry patch. The patch's whereabouts had been kept a family secret. After picking huckleberries there, I know why. It seemed like every bush held a cup of those delicious morsels. I doubt I will ever again see as productive a patch as that one. It is the only huckleberry patch I have ever seen where I picked a gallon of berries without too much trouble.

It pains me to think of all the practical knowledge, warmth, and love that was lost when Mrs. Moulton died. There would have been no greater tribute than to have listed her recipes in this book. I'm sorry to say I didn't record a one. I did want to take some space and thank her and her husband (John is now 103 years old and still lives in Jackson Hole) for sharing so much of their knowledge and love with us.

What I find amusing about huckleberries is that they aren't. You see, true huckleberries don't grow west of Minnesota. True huckleberries belong to the genus *Gaylussacia* and the berries we call huckleberries are actually blueberries and belong to the genus *Vaccinium*. This only affirms that when people talk of plants, common names are what they use. Since both genera belong to the same family, *Ericaceae*, there are few differences between them. I can honestly tell you that *wild* huckleberries and blueberries have the same flavor, so why worry what you call them? They both taste sensational, although the folks in Trout Creek contend that their huckleberries taste better than any huckleberry not from Montana's huckleberry capital!

For the sake of continuity, huckleberry is the name I'll stick with. Huckleberries and blueberries grow on acidic soil. Granite soils tend to be acidic, so look for huckleberries where you will find granite mountains. In Montana, the Bitterroot Mountains are a good place to start looking. In Wyoming, the Tetons are a great place, as are the ranges in Yellowstone National Park. Huckleberry is found from northern Wyoming west through Idaho and Oregon. The species (*Vaccinium globulosa*) does not occur in Colorado. Some of the related species like low bush blueberry and bilberry do occur there, however. According to Dr. Dee Strickler there are eight species of *Vaccinium* in Montana, so there are a good number of huckleberries, blueberries, bilberries, and whortleberries in the mountain West.

Huckleberries prefer moderate to full sunlight, so look for them in old burns or timber cuts. You'll not find them in dense shade, so avoid thick timber.

Huckleberries blossom May to July depending on the altitude where they grow. The blossoms look like little bells which are yellow-pink tinged to white. Huckleberries ripen early August through September. The ripe berry is dark purple. The berry has a scalloped ring around the tip.

Huckleberry ranges from a foot and a half to three feet tall and has brownish branches (as compared to low bush blueberry which seldom exceeds a foot and a half and has light green branches). The simple leaves are green, and two to three inches long. They have an oval or lance shape.

Huckleberries can be spotty. Sometimes you'll not get enough huckleberries to make a pie (three cups); other years you'll have enough for a batch of jam (two quarts) and several pies. Your chances are enhanced by knowing several patches of huckleberries. If you have three or four in mind, usually one of them will yield a decent picking of berries. In normal years I would rate your chances of finding two quarts or more at three out of four.

Huckleberries freeze well. Wash them, remove leaves and sticks, and drain in a colander. Place the berries on several thicknesses of paper toweling to dry thoroughly so the berries will not stick together when frozen. Place in one-quart freezer bags and pop into the freezer. You can even put one-cup portions in bags so you will have a premeasured amount to make huckleberry muffins, pancakes, or syrup on some winter's day.

Betty Bindl of Trout Creek, Montana, recommends that you pack the dried huckleberries "in fruit jars or other heavy freezer containers. Wash hands in hot, soapy water, and then wipe outsides of containers with a cloth dipped in the soapy water. The odor of the berry juice on the outsides of the containers will permeate other frozen food in the freezer if this step is omitted. Label containers, then freeze berries immediately."

Recipes

HUCKLEBERRY ROLYPOLY*

2 tablespoons, 1 cup sugar
2 cups Bisquick
⅔ cup, 2 cups water
1 cup huckleberries

Add 2 tablespoons sugar to 2 cups Bisquick and ⅔ cup water. Mix together. If sticky, add enough Bisquick so that it can be rolled. Roll the dough to a thin 9-inch wide rectangle. Spread 1 cup mountain huckleberries over the surface. Beginning at one 9-inch edge, roll the dough as for jelly rolls, being careful to keep berries spread evenly.

Meanwhile, bring to a boil 1 cup sugar and 2 cups water in a 9'' x 5'' metal bread pan. Place the rolled dough in the boiling syrup and bake in a 350 degree oven for 40 minutes or so—until it tests done in the middle with a straw.

Cool in the pan until just warm. Slice and serve in a bowl with cream.

Jean Young, Seattle, Washington

* This recipe is delicious with other berries, too—blueberries, service-berries, wild raspberries, or wild strawberries.

HUCKLEBERRY CHEESE PIE

1 8-ounce package cream cheese
1 cup thick sour cream
¾ cup, 2 tablespoons sugar
3 teaspoons vanilla
2 eggs, well-beaten
6 tablespoons water
2 tablespoons cornstarch
1 pint, 1 cup huckleberries
One pie shell, baked and cooled (see serviceberry recipes).

Mix together well: cream cheese, ½ cup sugar, eggs, and 2 teaspoons vanilla. Pour into crust and bake at 350 degrees for 35 minutes.

Blend sour cream, 1 teaspoon vanilla, and 2 tablespoons sugar. Spread this onto pie, at the end of cooking time above, to within ½ inch from edge. Place pie back into a hot oven, 425 degrees for 5 minutes. Cool.

Topping: Arrange about one pint huckleberries on top of cooled pie. Place one cup huckleberries in a saucepan with 3 tablespoons water and ¼ cup sugar. Simmer for 10 minutes and strain. Put juice back in saucepan (if juice does not equal 1 cup, add water) and add 2 tablespoons cornstarch mixed with 3 tablespoons water. Cook, stirring constantly until thickened. Cook 5 minutes more, stirring. Cool slightly, then pour over berries on pie.

Let pie set for at least one hour before cutting.

Top slices with a spoonful of sweetened heavy whipped cream.

Jeanette Wolf, Worden, Montana

HUCKLEBERRY CREAM CAKE

1½ cups heavy cream
2 teaspoons vanilla
3 eggs
1½ cups flour
1½ cups sugar
2 teaspoons baking powder
¼ teaspoon salt
1 cup huckleberries

Beat cream until stiff. Add vanilla. Beat eggs until thick and add to cream mixture. Sift together flour, sugar, baking powder, and salt and fold into cream mixture. Fold in huckleberries.

Put into greased and floured cake pans and bake for 30 minutes at 350 degrees. Cool.

Filling:

    1 cup huckleberries
    ¼ cup water
    ¼ cup sugar
    2 tablespoons cornstarch mixed with 1 tablespoon cold water

Simmer huckleberries, water, and sugar together for 10 minutes. Add cornstarch and water. Cook, stirring constantly, until thickened. Cool.

Frosting:
  Beat together:
    1 cup soft butter
    3 cups powdered sugar
    1 teaspoon vanilla
    4 egg yolks

Spread huckleberry filling between layers of cake, reserving about 2 tablespoons to garnish top of cake. Frost cake. Decorate top with reserved filling.

Jeanette Wolf, Worden, Montana

HUCKLEBERRY SOURDOUGH PANCAKES

2 cups flour
2 cups milk
1 teaspoon salt
1 cup sourdough starter

Beat together until smooth. Cover and let stand overnight. Beat down and save out 1 cup starter. To remainder, add:

2 eggs
3 tablespoons oil
2 teaspoons sugar
2 teaspoons baking soda

Mix well. Fold in ½ cup huckleberries. Let stand 5 minutes. Spread thinly on hot griddle. Turn once.

Judy Flugstad, Billings, Montana.
From the Billings Gazette Cookbook, *September 25, 1983.*

HUCKLEBERRY BON BONS

Chop (at least) 2 cups huckleberries. Set aside to drain. Should have about 1 cup.

1 8-ounce package cream cheese (softened to room temperature)
1 can lemon frosting mix
5-6 cups crumbs (vanilla cookie or graham crackers)
1 can white or chocolate frosting

Add huckleberries to all ingredients except white or chocolate frosting and mix well with fork. Chill in refrigerator for about an hour. Then form into 1-inch balls and return to refrigerator for 2 hours or overnight.

Dip the balls in white or chocolate frosting—heated in its own container set in a pan of water. Drain candies on wax paper to set. They freeze nicely.

Angela Loendorf, Wolf Point, Montana.
From the Billings Gazette Cookbook, *September 25, 1983.*

HUCKLEBERRY DESSERT

First layer:

  1 cup flour
  ½ cup butter, melted
  ¼ cup chopped walnuts or pecans

Blend like pastry and pat into 9'' x 12'' pan. Bake at 350 degrees for 10 to 15 minutes. Cool.

Second layer:

  1 8-ounce package cream cheese, softened
  1 cup powdered sugar
  1 9- or 10-ounce container Cool Whip

Beat soft cream cheese and powdered sugar until blended. Beat in Cool Whip. Spread over pastry.

Third layer:

  1 quart huckleberries
  ¾ cup water
  ¾ cup sugar
  2 teaspoons lemon juice
  5 tablespoons cornstarch

Heat berries, water, sugar and lemon juice. Dissolve cornstarch in ¼ cup cold water. Stir into huckleberries and cook until thickened and clear. Cool. Pour over second layer.

Fourth layer:

Top with 9-ounce container of Cool Whip. Sprinkle with nuts.

Darlene Ramsbacher, Fort Peck, Montana.
From the Billings Gazette Cookbook, *September 25, 1983.*

CANNED HUCKLEBERRIES

Wash huckleberries, picking over carefully. Fill clean jars with raw fruit. Shake down for a full pack. Add boiling water to ½ to 1 inch from jar top. Or, add boiling sugar syrup using the proportions of 2 or 3 cups sugar, (depending on sweetness desired) to 4 cups water. Adjust lids. Process in water bath 10 minutes for pints and 15 minutes for quarts.

Betty Bindl and Freida Park,
The Trout Creek Huckleberry Cookbook, *Trout Creek, Montana, 1984.*

HUCKLEBERRY CAKE

¼ cup shortening
1¼ cups sugar
2 eggs
1 teaspoon baking powder
2 cups flour
¼ cup milk
1½ teaspoons vanilla
1 to 1½ cups huckleberries

Cream the shortening and sugar, using a mixer. Beat in eggs. Mix dry ingredients together and add to the creamed mixture alternately with the milk. Blend well after each addition. Stir in vanilla, fold in berries. Pour batter into greased 9-inch square pan. Sprinkle crumb topping over batter. Bake at 350 degrees for 35-45 minutes.

Crumb topping: Mix together until crumbly ½ cup sugar, ½ cup flour, ½ teaspoon cinnamon, ¼ cup butter, melted.

Betty Bindl and Freida Park,
The Trout Creek Huckleberry Cookbook, *Trout Creek, Montana, 1984.*

QUICK HUCKLEBERRY COFFEE CAKE

1½ cups flour
¼ cup sugar
2½ teaspoons baking powder
¼ cup white shortening
1 egg
¾ cup milk
1 cup huckleberries

Combine dry ingredients in bowl; add shortening, egg and milk all at once. Beat 50 quick strokes with a fork. Stir in the huckleberries. Pour batter into a greased 9-inch square pan. Sprinkle top of batter with a mixture of:

½ cup sugar
½ teaspoon cinnamon
⅓ cup flour
¼ cup soft margarine

Bake 25 to 30 minutes at 375 degrees. Cut in squares.

Betty Bindl and Freida Park,
The Trout Creek Huckleberry Cookbook, *Trout Creek, Montana, 1984.*

HUCKLEBERRY SQUARES

½ cup margarine
1½ cups sugar
2 eggs
1 teaspoon vanilla
2 cups sifted flour
2 teaspoons baking powder
½ teaspoon salt
1 cup chopped walnuts
2 cups huckleberries

Melt margarine; remove from heat. Stir in sugar and cool slightly. Add eggs and vanilla; beat well. Combine dry ingredients; blend into egg mixture. Stir in nuts and huckleberries. Spread batter in greased 9'' x 13'' pan. Bake at 350 degrees for 35 minutes. Cut into 24 squares.

Betty Bindl and Freida Park,
The Trout Creek Huckleberry Cookbook, *Trout Creek, Montana, 1984.*

SOUR CREAM HUCKLEBERRY CAKE

1 cup sugar
1 cup sour cream
2 eggs
1 teaspoon vanilla
2 cups flour
1 teaspoon baking powder
1 teaspoon salt
2 cups clean, fresh huckleberries

Mix sugar, sour cream, eggs, and vanilla together. Add dry ingredients, mixing well. Gently fold in huckleberries. Bake at 350 degrees, about 30 minutes. Serve with cream.

Betty Bindl and Freida Park,
The Trout Creek Huckleberry Cookbook, *Trout Creek, Montana, 1984.*

HUCKLEBERRY COOKIES

½ cup butter
1 cup sugar
1 teaspoon lemon juice
½ teaspoon grated lemon rind
1 egg
2¼ cups sifted flour
2 teaspoons baking powder
1 teaspoon cinnamon
¼ teaspoon nutmeg
¼ cup milk
1 cup grated carrot
1 cup huckleberries

Cream butter and sugar until light and fluffy. Add lemon juice, rind, and egg; beat until well-blended. Sift flour, baking powder, cinnamon, and nutmeg. Add to creamed mixture alternately with milk. Stir in carrot and huckleberries. Drop by rounded teaspoonsful on greased cookie sheet. Bake at 375 degrees for 12 to 15 minutes or until done. Yield: about 4 dozen.

Betty Bindl and Freida Park,
The Trout Creek Huckleberry Cookbook, *Trout Creek, Montana, 1984.*

HUCKLEBERRY SOUR CREAM PIE

2 eggs, slightly beaten
¾ cup sugar
¼ teaspoon salt
1 teaspoon cinnamon
½ teaspoon nutmeg
¼ teaspoon cloves
1 cup sour cream
1 cup fresh or frozen huckleberries
Pastry for double crust pie (see serviceberry recipes)

Preheat oven to 425 degrees. Prepare pie crust in 8-inch pie plate. In saucepan, mix slightly beaten eggs, sugar, salt, cinnamon, nutmeg, cloves, and sour cream. Cook on medium heat, stirring constantly, until thick. Remove from heat and fold in huckleberries. Pour into pastry-lined pan and cover with top crust. (Cut air vents in top crust). Bake at 425 degrees for 10 minutes, then at 350 for 20 minutes.

Betty Bindl and Freida Park,
The Trout Creek Huckleberry Cookbook, *Trout Creek, Montana, 1984.*

HUCKLEBERRY DREAM

2 pastry shells (see serviceberry recipes)
toasted pecans
1 large package Dream Whip
1 cup cold milk
2 cups powdered sugar
1 8-ounce package cream cheese
1½ cups huckleberry pie filling

Place pecans in bottom of pie shells. Whip Dream Whip with milk until stiff. Add sugar. Stir in softened cream cheese. Place in pie shells over toasted pecans. Spread huckleberry pie filling over top. Chill. Makes two pies.

Betty Bindl and Freida Park,
The Trout Creek Huckleberry Cookbook, *Trout Creek, Montana, 1984.*

HUCKLEBERRY TARTS

2 tablespoons cornstarch
½ cup powdered sugar
2 cups huckleberries, divided
½ cup water
2 tablespoons lime juice
1 cup whipping cream
8 3-inch baked tart shells

Combine cornstarch and sugar, stir in ½ cup huckleberries, water, and lime juice. Cook, stirring over medium heat until mixture is clear and thickened. Cool. Whip cream, fold into cooked mixture along with remaining berries. Spoon into tart shells. Refrigerate. At serving time, garnish with additional whipped cream, if desired. Yield 8 servings.

Betty Bindl and Freida Park,
The Trout Creek Huckleberry Cookbook, *Trout Creek, Montana, 1984.*

POOR MAN'S HUCKLEBERRY PIE

8-inch pastry shell (see serviceberry recipes)
2 baking powder biscuits, crumbled
½ cup granulated sugar
½ teaspoon cinnamon
dash of nutmeg
1 cup light cream
1 cup fresh huckleberries
½ cup chopped nuts

Use leftover biscuits. Break or crumble the biscuits over the unbaked pastry crust. Sprinkle with the sugar, cinnamon, and nutmeg. Pour cream over biscuits and top with nuts and huckleberries. Bake in 350 degree oven for 40 to 50 minutes, until knife inserted in center comes out clean. Serve warm.

Betty Bindl and Freida Park,
The Trout Creek Huckleberry Cookbook, *Trout Creek, Montana, 1984.*

BARELY BERRY AND CUSTARD PIE

Pie crust dough (see serviceberry recipes)
1 cup fresh or frozen huckleberries
1 cup milk
1 egg, beaten
¾ cup sugar
½ teaspoon vanilla
dash salt
1 tablespoon butter or margarine
1 teaspoon cinnamon

Roll out pie dough to fit 10" x 15" pan, with 1-inch crust overhanging edge. Turn under to form a holding edge.

Sprinkle on berries. Mix together milk, egg, sugar, vanilla, and salt. Pour over berries. Dot with butter, sprinkle on cinnamon. Bake 1 hour at 350 degrees. Serve in squares. If times aren't too hard, can be topped with whipped cream.

Betty Bindl and Freida Park,
The Trout Creek Huckleberry Cookbook, *Trout Creek, Montana, 1984.*

HOMESTEAD HUCKLEBERRY PUDDING

¼ cup shortening
½ cup sugar
1 cup flour
¼ teaspoon salt
2 teaspoons baking powder
½ cup milk
2 cups huckleberries
¾ cup sugar
¾ cup boiling fruit juice or water

Cream ½ cup sugar together with shortening. Mix in the dry ingredients alternately with milk. Pour into a 2-quart casserole dish. Mix berries, ¾ cup sugar, and water together and pour over the top of the batter. Bake at 350 degrees, until the batter comes to the top, about 50 minutes. Serve plain or with cream.

Betty Bindl and Freida Park,
The Trout Creek Huckleberry Cookbook, *Trout Creek, Montana, 1984.*

HUCKLEBERRY SHERBET

2 cups water
1 cup sugar
1½ cups mashed huckleberries
1 tablespoon lemon juice
1 egg white

Boil water and sugar 5 minutes. Cool and add berry pulp that has been rubbed through a coarse sieve. Add lemon juice. Pack in ice and salt and freeze to a mush. Add stiffly beaten egg white and finish freezing.

Betty Bindl and Freida Park,
The Trout Creek Huckleberry Cookbook, *Trout Creek, Montana, 1984.*

HUCKLEBERRY CRISP

2 cups huckleberries
½ cup sugar
¼ cup water
⅛ teaspoon salt
1 tablespoon lemon juice
2 cups corn flakes
¼ teaspoon cinnamon
2 tablespoons melted butter

Rub a 1½ quart baking dish with butter. Combine berries, sugar, water, salt, and lemon juice in saucepan and simmer 5 minutes. Pour half the berry mixture into baking dish. Top with 1 cup coarsely crushed corn flakes. Repeat. Sprinkle with cinnamon, drizzle butter over top. Bake 25 minutes at 375 degrees. Serve with cream.

Betty Bindl and Freida Park,
The Trout Creek Huckleberry Cookbook, *Trout Creek, Montana, 1984.*

HUCKLEBERRY BUCKLE

Sift together in a bowl:
    ¼ cup sugar
    2½ teaspoons baking powder
    2 cups flour
    ¼ teaspoon salt

Make a well and add the following:
    1 egg
    ¼ cup melted fat
    ½ cup milk

Beat liquid ingredients, then slowly stir just enough to mix well with the flour. Pour into a shallow glass baking dish. Cover with 1 pint huckleberries. Top with the following crumb mixture:

¼ cup sugar
⅓ cup flour
¼ cup butter
½ teaspoon cinnamon

Bake in a 13'' x 9'' baking dish for 40-50 minutes at 350 degrees. Yield: 12 three-inch squares.

Betty Bindl and Freida Park,
The Trout Creek Huckleberry Cookbook, *Trout Creek, Montana, 1984.*

HUCKLEBERRY CORN MUFFINS*

1¼ cups sifted flour
¾ cup yellow cornmeal
2 to 4 tablespoons sugar
4½ teaspoons baking powder
1 cup huckleberries, sugared
1 teaspoon salt
1 egg
⅔ cup milk
⅓ cup melted margarine or oil

Heat oven to 400 degrees. Into medium bowl, sift flour, cornmeal, sugar, baking powder, and salt. In small bowl, beat egg well with fork, stir in milk, butter; pour all at once into flour mixture, stirring with fork until flour is just moistened. Fold in berries. Fill 16 greased muffin cups two-thirds full. Bake 25 to 30 minutes, or until done. Serve hot.

Betty Bindl and Freida Park,
The Trout Creek Huckleberry Cookbook, *Trout Creek, Montana, 1984.*

*Serviceberries, low bush blueberries, or wild strawberries can be substituted.

HUCKLEBERRY NUT BREAD

2 cups sifted flour
3 teaspoons baking powder
¼ teaspoon salt
1 cup sugar
1 cup huckleberries
½ cup chopped nuts
2 well-beaten eggs
1 cup milk
3 tablespoons salad oil

Sift flour, baking powder, salt, and sugar. Add berries and nuts. Add eggs mixed with milk and oil. Stir slightly. Pour into paper-lined 9" x 5" x 3" loaf pan. Let stand 30 minutes. Bake in moderate oven (350) for 1 hour.

Betty Bindl and Freida Park,
The Trout Creek Huckleberry Cookbook, *Trout Creek, Montana, 1984.*

HUCKLEBERRY WINE

1 gallon huckleberries
1 gallon water
6 to 8 cups sugar (do not cut sugar—it is needed in fermentation process)

First day, put berries in a crock (not metal) and mash. Heat ½ gallon water to boiling and pour over the berries.

Second day, let mixture sit.

Third day, dissolve sugar in another ½ gallon boiling water. Let sugar water cool to room temperature and add to berry and water mixture.

Every day for two weeks, stir the mash. Then strain through cheesecloth and set aside pulp, keeping juice. Return juice to warmed receptacle. Let settle for three days; siphon juice into bottle and cork lightly.

Betty Bindl and Freida Park,
The Trout Creek Huckleberry Cookbook, *Trout Creek, Montana, 1984.*

HUCKLEBERRY N.Y. CAKE

Preheat oven to 350 degrees.
Grease and flour 2 loaf pans.
Cream together 1 cup butter or margarine and 2¼ cups sugar.
Add 4 eggs and 1 teaspoon vanilla—beat until smooth.

Sift together: 4 cups flour, 1 teaspoon salt, and ½ teaspoon soda—add alternately with 1 cup sour milk (add 1 tablespoon vinegar to 1 cup sweet milk).

Add to batter and blend 1½ cups huckleberries.

Bake at 350 degrees for approximately 1 hour.

Huckleberry Recipes,
Compiled by the Swan Lake Women's Club, Swan Lake, Montana.

HUCKLEBERRY WHIRL SALAD

3 small packages black cherry Jello
2 cups boiling water
5 ice cubes
2 cups miniature marshmallows
1 medium can crushed pineapple and juice
2 cups huckleberries
2 cups pecans
3 tablespoons mayonnaise

Dissolve Jello in boiling water, add marshmallows, and stir until melted. Add ice cubes, stirring until melted. Let thicken, then add pineapple, huckleberries, nuts, and mayonnaise—mix well. Place in mold and let set in refrigerator.

Huckleberry Recipes,
Compiled by the Swan Lake Women's Club, Swan Lake, Montana.

HUCKLEBERRY-MINT JELLY

9 cups huckleberries
1 cup fresh, wild mint leaves
1½ cups of water
¾ cup sugar per cup of juice
1 pouch Certo liquid pectin

Wash and clean the huckleberries. Chop in a blender. Place in a 6- or 8-quart saucepan or kettle and add ½ cup mint infusion.

Mint infusion: crush 1 cup of fresh, wild mint leaves. Place in separate saucepan and bring to a boil. Cover and allow to steep for 10 minutes. Strain the liquid through cheesecloth. Measure ½ cup of the mint solution.

Simmer mint infusion and berry mixture for five minutes. Remove from heat and strain the mixture through cheesecloth or a jelly bag. Measure the juice and add ¾ cup of sugar to each cup of juice. Stir until sugar dissolves. Place over heat and bring to a boil. Add liquid pectin and bring to a full, rolling boil. Boil for one minute. Remove from heat, skim off foam, and pour into sterilized jars. Seal and process in boiling water bath for 10 minutes. Yield: six 8-ounce jars.

Cel Hope, Sheridan, Wyoming

Low Bush Blueberry

Low bush blueberry permeates lodgepole pine and spruce-fir stands throughout the mountain West. The short, pale green shrub seldom bears fruit in abundance in successive years.

In good years, low bush blueberry will color the forest floor a dull, purplish red. In other years, the same stands will be a monotonous, pale green—the bushes will have a total absence of berries due to late frosts killing the blossoms and eliminating the berry crop.

In some localities, people call low bush blueberries grouse whortleberries, or whortleberries. Whatever, they are a common inhabitant of the pine and spruce forests of the mountain West.

Low bush blueberries aren't big berries to begin with—not much bigger than a BB—but they are one of the sweetest, tastiest berries of the West. When they are bountiful, they are easily attainable by hikers. The berries make great trail snacks. Though the small size of the berry discourages a person from picking large amounts, its excellent flavor makes a person keep picking more.

Low bush blueberries are oh-so-good eaten as is, fantastic with a touch of sugar and milk, and out of this world in pancakes and syrup.

I don't know how many times low bush blueberries have saved my hiking trips, but I daresay that the number is several dozen. I always pack enough freeze-dried food to cover meals, but I am always hungry on a hike and it often seems that when I have a real craving for some fresh fruit, I stumble onto a patch of low bush blueberries. Sometimes the berries are so thick that I can strip one branch at a time and come up with a handful in two or three swipes. Needless to say, after three or four handfuls, I am ready for the trail again. Sometimes I even pick a half-cup of the berries to take along for my dinner dessert, or breakfast addition.

I once thought that it would be impossible to pick enough low bush blueberries to make anything substantial like jam. An outfitter from Story, Wyoming, Bob Szewc, set me straight on the matter. He says that his wife Jeanne figured out how to harvest them in decent quantities. She uses a big-toothed comb and a large bowl to literally comb the bushes. Where I would break my back to pick a cup or so, she can get quarts of low bush blueberries in an equal amount of time. Consequently, she has enough berries for jams, muffins, pies, and cobblers and maybe even some left over to freeze for winter uses.

Low bush blueberry seldom exceeds a foot in height. Its scientific name, *Vaccinium scoparium*, puts it in the same genus as cranberries and bilberries. The leaves are a light, glossy green color and less than a half-inch in length. The leaves are very thin and have a serrate margin.

The stems are pale green color and angled. Many times the branches are so thick that the bush resembles a broom.

Low bush blueberry flowers May through June depending on elevation and exposure. The white or pink flower is bell- or urn-shaped. The berry ripens as early as late July or as late as mid-September. The BB-sized berry is red to red-purple in color.

Your chances of picking a quart of low bush blueberries are about one in ten most years. Your chances of getting a cup are about four in five.

Low bush blueberries freeze well. Wash them and let them drain. Place the berries in plastic freezer bags and put them in your freezer. They should be good for six months or so. 🍓

Recipes

As I mentioned in the text, low bush blueberries are great for trail snacks. I find that they are good eaten as is, with a little milk and sugar, or in muffins or pancakes.

BLUEBERRY PANCAKES

This is a trail recipe. Mix up a batch of pancake batter (I like the pancake mix that you only have to add water for hiking expeditions), stir in ¼ cup of blueberries and cook them up as you would normal pancakes. I find that I have to use a tad more cooking oil to keep the pancakes from sticking.

BLUEBERRY SYRUP

Take ¼ cup (a metal hiker's cup, that is) of low bush blueberries, crush with a fork or spoon, add two tablespoons of water and two tablespoons brown sugar and simmer for ten to fifteen minutes. This syrup is a perfect complement to the above pancakes.

IMPOSSIBLE CHEESECAKE*

¾ cup sugar
2 eggs
2 teaspoons vanilla
½ teaspoon grated lemon peel
½ cup Bisquick
2 8-ounce packages cream cheese, softened and cut into cubes

Place all ingredients into a blender. Blend on high for 3 minutes. Pour into greased 9" x 1¼" pieplate and bake at 350 degrees for 30 minutes, or just till puffed and center is dry. Spread cheesecake topping over top and cool. Chill 3 hours. Serve with fresh low bush blueberries (about 1 cup) sprinkled over top.

Cheesecake topping:
    Mix 1 cup dairy sour cream, 2 tablespoons sugar, and 1 teaspoon vanilla.

*Strawberries, raspberries, blackberries, huckleberries, or juneberries may be used.

Charlotte Heron, Missoula, Montana

WHORTLEBERRY MUFFINS

2 cups flour
2 teaspoons baking powder
2 eggs
3 tablespoons oil or melted butter
½ teaspoon salt
⅓ - ½ cup of sugar
¾ cup milk
1 cup whortleberries, lightly floured

Mix dry ingredients. In a separate bowl, beat eggs, add milk and oil. Pour wet ingredients into dry ingredients and mix gently. Don't overmix. Fold in berries. Pour into greased muffin pans and bake at 400 degrees for 20-25 minutes. Makes 2 dozen small or 1 dozen large muffins.

"I use this for huckleberries, too. The secret is to flour the berries and don't overmix." Charlotte Heron, Missoula, Montana.

SOUR CREAM BLUEBERRY PANCAKES*

1 cup flour
¼ teaspoon salt
3 teaspoons baking powder
1 tablespoon sugar
1 egg
1 cup milk
¼ cup sour cream
2 tablespoons melted butter
½ cup low bush blueberries

Sift dry ingredients together, beat together egg, milk, and sour cream. Pour milk mixture over dry ingredients, blend until smooth. Add melted butter, mix well, fold in blueberries. Bake on hot griddle.

*Huckleberries, serviceberries, and wild strawberries can also be used.

Huckleberry Recipes,
Compiled by the Swan Lake Women's Club, Swan Lake, Montana.

WHORTLEBERRY COBBLER

1 pint whortleberries
⅓ cup water
1½ cups raw sugar
1 teaspoon grated lemon rind
1 cup whole wheat flour
½ teaspoon sea salt
1 teaspoon baking powder
⅓ cup butter

Preheat oven to 350 degrees.

Combine the whortleberries, water, ¾ cup of the sugar, and the lemon rind in a heatproof casserole. Bring to a boil and simmer two minutes. Meanwhile, combine remaining sugar with flour, salt, and baking powder. Cut in the butter until mixture is crumbly. Sprinkle crumbs over fruit. Bake about 25 minutes or until browned. Serve warm. Yield: 4 servings.

Huckleberry Recipes,
Compiled by the Swan Lake Women's Club, Swan Lake, Montana.

BLUEBERRY CHEESECAKE

Make 1 (9'') graham cracker crust. Bake.

1 8-ounce package cream cheese, softened
⅓ cup lemon juice
1 teaspoon vanilla
1 14-ounce can sweetened condensed milk

Blend ingredients above and pour into baked crust. Chill.

Glaze:
    2 cups blueberries
    3 tablespoons cornstarch
    1 cup sugar
    ½ cup water or berry juice

Cook over medium heat till thick. Cool and pour over cool cheesecake. Chill thoroughly.

Huckleberry Recipes,
Compiled by the Swan Lake Women's Club, Swan Lake, Montana.

BLUEBERRY BUCKLE

½ cup margarine
1 cup sugar
1 egg, beaten
¼ teaspoon salt
2 cups flour
2½ teaspoons baking powder
½ cup milk
2 cups blueberries—frozen berries may be used if thawed.

Cream margarine and sugar. Add egg and mix well. Sift flour, salt, and baking powder and add alternately with milk. Fold in berries (makes a stiff dough). Place in a greased 8-inch pan. Sprinkle with the topping.

Topping:
    ½ cup sugar
    ½ teaspoon cinnamon
    ¼ cup soft margarine
    ½ cup flour

Mix ingredients with fork.

Bake at 350 degrees, 1 hour 15 minutes.

Huckleberry Recipes,
Compiled by the Swan Lake Women's Club, Swan Lake, Montana.

PEACH-BLUEBERRY PIE

2 tablespoons lemon juice
3 cups peaches (sliced)
1 cup blueberries
1 cup sugar
1 tablespoon tapioca
½ teaspoon salt

Mix all ingredients in a bowl and set aside while making crust for double-crust pie.* Pour filling in prepared unbaked crust and top with 2 tablespoons butter and upper crust. Bake at 400 degrees until pie is brown as you desire.

*See pie crust recipe in serviceberry chapter.

Huckleberry Recipes,
Compiled by the Swan Lake Women's Club, Swan Lake, Montana.

Oregon Grape

Oregon grape is anything but a grape, though the ripe fruits look like small grapes. Oregon grape belongs to the barberry family *Berberidaceae* whose relatives include barberry and holly.

While Oregon grape may be misnamed, it is an interesting plant. It has other uses than just jelly; for instance, its roots make an excellent yellow dye.

Oregon grape loves pine forests and rocky foothills. It seems to have a strong affinity for ponderosa and (to a lesser degree) lodgepole pine. In other words, you will probably find Oregon grape wherever you find a stand of ponderosa.

It is one of those berries I pick incidentally to another activity, like when I'm just wandering through the foothills. I usually don't search for Oregon grape, but if I find it, I pick it. It seems that I find a lot of different berries that way.

Since Oregon grape ripens in early autumn, I can usually take my dogs, Midi and Millie, and search the foothills for ruffed and blue grouse. It is always handy to take a plastic bag along in case I find a good supply of Oregon grape while I'm sauntering about the hills. Usually I end up with more grapes than birds, and that's just fine by me, for I do enjoy this berry. It makes splendid jelly!

Oregon grape belongs to the genus *Mahonia*. The common, low-growing species of the region is *Mahonia repens*. Tall Oregon grape (*M. aquilifolium*) is native west of the Cascade Mountains, but not to the Rockies. It is the state flower of Oregon. It has been planted extensively in other parts of the country, including western Montana, so you might find some escapes there and adjacent to cities and towns elsewhere.

The hardy low-growing species blooms early—late April into May. It seems to blossom two to three weeks after the snows leave the foothills.

The blossoms are bright yellow clusters (racemes).

Oregon grape has pinnately compound leaves. The leaflets look a lot like holly leaves, complete with spines on the edges. If Oregon grape grows in a sunny area, the leaves will be red-tinged; those growing in the shade will have a deep green color. The leaflets are rather thick and leathery.

A short plant seldom exceeding a foot in height, Oregon grape has underground stems that enable it to spread. The grapes mature in September, and are dark blue with a powdery blush. The grapes will be in clusters of three to ten and be about the size of peas.

While it is most common in western Montana, you can find pickable patches in Wyoming, Colorado, Utah, and Idaho.

Your chance of getting enough Oregon grape for a batch of jelly (two quarts plus) is about one in four.

Recipes

OREGON GRAPE JELLY

3½ pounds Oregon grapes
1 cup water
¼ cup lemon juice

Wash and crush Oregon grapes. Add 1 cup of water and ¼ cup lemon juice. Simmer, covered, until juice is free, 15-20 minutes. Squeeze out juice.

3½ cups juice
1 package MCP pectin
4½ cups sugar

Add juice to a kettle, stir in pectin until it dissolves. Place over high heat; bring to boil, stirring constantly to avoid scorching.

Add measured sugar; mix well. Continue stirring and bring to a full rolling boil (a boil that cannot be stirred down). Boil hard exactly 2 minutes. Remove from heat. Skim foam and pour into hot, sterilized glasses to within ⅛-inch from top. Place lids on jars; tighten bands and place in a hot water bath for 10 minutes.

Charlotte Heron, Missoula, Montana

OREGON GRAPE-APPLE JELLY

2 cups Oregon grape juice
1½ cups apple juice concentrate
1 package MCP pectin
4½ cups sugar

Add juices to a kettle, stir in pectin until it dissolves. Place over high heat; bring to boil, stirring constantly to avoid scorching.

Add measured sugar; mix well. Continue stirring and bring to a full rolling boil (a boil that cannot be stirred down). Boil hard exactly 2 minutes. Remove from heat. Skim foam and pour into hot, sterilized glasses to within ⅛-inch from top. Place lids on jars; tighten bands and place in a hot water bath and boil for 10 minutes.

"Oregon grape jelly is rather tart. This recipe lessens the intensity of the flavor." Charlotte Heron, Missoula, Montana.

Serviceberry

Serviceberry goes by several common names in the West. In eastern Montana, many folks refer to it as juneberry. In Jackson Hole, people call it "sarvisberry." Whatever you call the berry, it has a first-class taste.

I once had the misconception that the only thing you made from serviceberry was jelly. Boy, was I wrong!

I can recollect calling my cooking-expert friend, Evelyn Hejde. She and her husband, Chester, live on a ranch near Aladdin, which is in the Wyoming portion of the Black Hills northeast of Sundance. I was looking for a recipe for serviceberry and asked Evelyn what she did with serviceberry besides make jelly.

She said, "I never have made jelly out of serviceberry, but lots of pies."

A few months later, I was talking with a lady from Hardin, Montana, Alice Powell. We were admiring a fine juneberry bush in Alice's yard and I asked if she made jelly out of the juneberries. She said, "No, they never last that long. We either have them with milk and sugar or with our cold cereal for breakfast. We never let enough accumulate to do anything major with them."

I did some reading and found out that serviceberry was one of the primary ingredients in pemmican—an Indian winter staple consisting of berries, meat, and fat.

Just as I thought there was only one use for serviceberry, I also believed there was only one way to pick them—one at a time by hand. My friend Cliff Schultz told me that when he was a kid in western North Dakota his parents picked serviceberries differently. They would tie towels to the ends of the serviceberry branches and to the trunks of the shrubs. Then they would shake the bush. The ripe berries would tumble into the towels, the unripe ones would remain on the bush. That sounds like a real labor-saving technique to me.

Serviceberry is quite well represented in the Rocky Mountain West. There are three species occurring here: *Amelanchier alnifolia, A. utahensis, A. pumila*. The differences between the species are slight. They occur from the subalpine regions to the foothills and into the brushy draws of the higher plains throughout the region. The most common one in the region is Saskatoon serviceberry, *A. alnifolia.*

Serviceberry is called shadbush in the northeast. It was named such because it blossomed in the early spring just as the shad came to spawn in the freshwater streams. In the Rocky Mountain region, serviceberry blossoms about the same time as wild plum, a little before chokecherry, and well before wild strawberry or wild raspberry. Depending on the elevation and latitude, serviceberry will blossom late April through late May (say for the foothills in Jackson Hole). One of the most beautiful spring scenes I have ever witnessed was the serviceberry blossoming in Clark's Fork Canyon just east of Missoula in early May, 1989. It seemed like the lower slopes of the canyon were a giant, white patchwork quilt.

Serviceberry blossoms occur in clusters called racemes. A raceme is defined as an elongated inflorescence (flower cluster) with a single main axis along which single, stalked flowers are arranged. Chokecherry and Oregon grape have racemes, too.

Serviceberry grows to nearly twenty feet in height, though ten to twelve feet is more common. The taller bushes make for difficult picking. I know that my good buddy Dick Appel and I had to struggle to pick bushes one July day a few years ago. Dick finally managed to pull the branches down while I picked like crazy. It seems that serviceberry picking can be a two-person undertaking.

Serviceberry ripens mid-July to mid-August. The berry turns from green to fuchsia to deep purple. Botanists call the berry a "pome." Serviceberry, apple, and hawthorn all are pomes. The size of serviceberry pomes ranges from pea-sized to two-and-a-half times that size. Serviceberry will have a star-shaped arrangement of sepals at the tip of the berry—just like an apple.

The leaf is rounded and has teeth from the tip to about midway on the margin. The leaf is dark green on top and paler underneath. Many times the underside of the leaf is hairy.

Serviceberry branches are quite smooth and have a reddish brown or light gray color.

Serviceberry is a highly preferred food for birds and mammals. Many times the birds harvest the ripe serviceberries before berry pickers can get to them. Deer, elk, and moose browse the branches quite heavily during the winter.

Serviceberry is sometimes afflicted with a rust infestation that can cause the berries to have contorted growth. Sometimes insects infest serviceberry and cause them to drop before they are ripe.

Since serviceberry has a host of maladies, a berry picker cannot be too confident of finding enough berries to pick. Most seasons I would rate your chances of finding two quarts at about fifty-fifty.

If you happen to get a surplus of serviceberries, you can store them easily in the freezer. Just put them in quart- or gallon-sized freezer bags and toss them in. They will keep well for six months or so—ample time to make them into jelly, pies, or muffins. 🐿

Recipes

SERVICEBERRY (JUNEBERRY) PIE

3 1/2 cups serviceberries
3/4 cup sugar
1 teaspoon cinnamon
2 tablespoons flour
1 tablespoon lemon juice
Pastry for a two-crust pie (see recipe in this chapter)

Mix all the above ingredients together, coating the serviceberries well. Put in 9-inch pie shell and cover with top crust and flute edges. Bake at 375 degrees 1 hour.

Marlene Welliever, Wibaux, Montana

Buffaloberry
blossoms are
miniscule.

Ripe buffaloberries
are either red or
orange. The leaves
have a rounded end
and silvery green
undersides.

Black currant
blossoms have a
trumpet shape and
spicy fragrance.

Ripe, ripening, and
green currants. Note
the ''pigtails'' on
the berries and the
maple-like leaves.

Chokecherry blossoms.
This type of an
inflorescence is called a
raceme.

Ripe chokecherries are
glossy black.

Elderberry blossoms
in June. The floral
arrangement is
called a ''cyme.''
With blue elderberry
the cyme is flat;
black elderberry has
a convex cyme.

Ripe blue elderberries.

Gooseberry has a trumpet-shaped blossom.

Ripe and ripening gooseberries. Like currants, gooseberries have pigtails.

Hawthorn blossoms resemble apple blossoms.

Ripe haws look like miniature apples.

Huckleberry
blossoms look like
miniature bells.

Ripe huckleberries
have a scalloped
edge.

Low bush
blueberries have
bell-shaped flowers.

A loaded low bush
blueberry bush.

Oregon grape blossoms early in the spring. The plant doesn't grow higher than six inches. Photo by Mike Aderhold.

Ripe Oregon grapes resemble wild grapes.

Serviceberry blossoms occur in racemes like chokecherries.

Ripe and ripening serviceberries. Note the ripening berry is a bright pink and that the leaf is dark green on top and light green underneath. The leaf has serrations on part of the margin.

Wild grape blossoms are quite small and nondescript. The leaves have large teeth and shallow lobes.

Ripe grapes are a deep blue color with a powdery blush.

Wild plum blossoms often paint the early spring landscape.

Ripe plums have a deep pink color with a blush. When they occur as thickly as this, the picking is easy.

Wild raspberries have rather nondescript blossoms with pale white petals. Note the prickly canes.

Ripe and unripe raspberries.

The blossom of a wild rose is not only beautiful, but fragrant.

Rose hips ripen in September. Note the stringy sepals that are retained at the tip of the hip.
Photo by Donald A. Schreder

Wild strawberry blossoms have five white petals. Note the lack of woody parts on this plant. Photo by Jerry Pavia

Ripe wild strawberries are not very big, but the flavor is king-size!

POISONOUS Nightshade fruits look like small Italian tomatoes. The fruits contain solanine and can be poisonous to children.

POISONOUS Baneberry has glossy red or white berries. The plant grows in moist areas in the mountains. The berries and roots are poisonous and can cause death.

JUNEBERRY TOPPING

1 pint canned juneberries (see page 105)
Drain juice into measuring cup.
Reserve one cup or add water to make 1 cup liquid.
1 tablespoon cornstarch

Stir together cornstarch and berry juice and heat until thick. Add the berries and stir until well-mixed. Spoon over ice cream or un-iced vanilla or yellow cake squares.

Barb Griffith, Helena, Montana

SERVICEBERRY MUFFINS

2 cups flour
1 teaspoon salt (scant)
1/4 cup sugar
3 teaspoons baking powder
1 cup milk
1 well-beaten egg
1/3 cup oil
1 cup berries, canned, frozen, or fresh (if canned, drain well).

Combine dry ingredients. Beat egg in good-sized bowl. Add the milk, oil, and berries. Stir gently to preserve whole berries. Add the combined dry ingredients. Stir just to moisten. Do not overmix.

Place in greased muffin tins and bake at 400 degrees for 20 minutes. Makes 12 muffins.

Barb Griffith, Helena, Montana.

SERVICEBERRY MUFFINS

1 cup serviceberries
3/4 cup sugar
2 1/2 cups flour
1/2 teaspoon salt
1 teaspoon baking powder
1 teaspoon baking soda
1 egg
3/4 cup buttermilk
1/4 cup cooking oil

Combine berries with sugar. In separate bowl, sift flour and add all other dry ingredients. In another bowl, beat egg and add buttermilk and oil. Add liquid ingredients to dry ingredients. Mix lightly. Dough will be lumpy. Fold in berries and sugar. Fill well-greased muffin pans 2/3 full. Bake at 400 degrees for 20-25 minutes.

Theo Hugs, Ft. Smith, Montana

SERVICEBERRY CONSERVE

4 cups chopped serviceberries
water
3 cups sugar
2 lemons
2 oranges

Chop the serviceberries. Place in a large saucepan or kettle and add enough water to cover the chopped berries. Cook until tender. Add 3 cups of sugar, the juice of two lemons, the shredded pulp of two oranges, and the juice. Grate the orange rinds and add it to the mixture. Heat to boiling, then lower heat and bubble for one half hour. Pour into sterile eight-ounce jars, seal, and process in a boiling water bath for ten minutes. Yield: four to five 8-ounce jars.

Joan Dixon, Sheridan, Wyoming

EVELYN'S BEST PIE CRUST RECIPE

4 cups flour
1 tablespoon sugar
1 teaspoon salt
1 3/4 cups shortening
1/2 cup water
1 egg
1 tablespoon vinegar

Combine dry ingredients. Cut in shortening until it is thoroughly mixed. Combine water, egg, and vinegar, stir with a fork and add to flour mixture. Mix thoroughly. Place in refrigerator and chill for at least 15 minutes. Roll out on a well-floured surface. Yield: Two 2-crust pies.

Evelyn Hejde, Aladdin, Wyoming

HERB-SPICED SERVICEBERRIES

15 cups of ripe serviceberries
3 pounds sugar
1 cup vinegar
1 cup of water

Spices:
6 whole cloves
1 tablespoon whole allspice
1 stick whole cinnamon
Selected herb (tarragon, basil, or mint)

Make a syrup of 3 pounds sugar, 1 cup vinegar, and 1 cup water. Add spice bag, bring to a boil, and cook for 10 minutes; remove from heat and allow to cool.

Add washed, stemmed serviceberries and heat to simmering. Allow to simmer until berries soften. Remove from the heat, cover quickly, and let stand overnight.

Next day remove spice bag and spoon the fruit into hot sterile jars. Reheat the syrup to a boil and pass selected herb through the hot syrup until you obtain the desired flavor and aroma.

Pour the herb syrup over the berries, seal the jars. Process in boiling water bath for 10 minutes. Allow the spiced berries to sit for 30 days before serving. This will give the spicy herb syrup a chance to do its work.

Cel Hope, Sheridan, Wyoming

SERVICEBERRY JELLY

3 1/3 cups serviceberry juice
1 package powdered pectin
5 cups sugar
1/4 cup lemon juice (optional)

Pick over and wash the serviceberries. Place the berries in a large kettle and barely cover with water. Bring to a boil, then reduce the heat so that the mixture barely boils. Cook for about 10 minutes or until there is a deep-colored liquid. The berries can be crushed as they cook or the first juice can be drained into another kettle and the berries can be cooked a second time. Crush the berries as they cook the second time to release more juice.

Strain the cooked serviceberries through a jelly bag or three thicknesses of cheesecloth.

Measure the juice accurately into a large (4-quart) pot or kettle. Add powdered pectin (and lemon juice) and stir to dissolve, then bring to a quick hard boil over high heat, stirring occasionally. Add premeasured sugar all at once. Bring to a full rolling boil (a boil that cannot be stirred down). Boil hard for one minute, stirring constantly.

Pour into hot, sterilized jelly jars. Fill to 1/8 inch of top. Wipe rim clean, place hot metal lid on jar, and screw metal band down firmly. Process in boiling water bath for 5-10 minutes.

Margaret and Charles Butterfield,
Preserving Wyoming's Wild Berries and Fruits,
University of Wyoming Agricultural Extension Service, 1981.

PEMMICAN

3 cups jerky, ground into powder
3 cups dried huckleberries or serviceberries
3 cups raw sunflower seeds or chopped raw peanuts
2 tablespoons raw honey or molasses
3/4 cup peanut butter
1 teaspoon freshly ground black pepper, optional
10 or 12 crushed juniper berries, optional

Mix ground jerky with berries and nuts. Mix honey, peanut butter, and pepper over low heat till well blended. Blend all ingredients together. Press into sausage casings and store hanging in cool, dry place, or freeze. Makes about 10 cups.

Patricia Mahana, Billings, Montana.
From the Billings Gazette Cookbook, *September 25, 1983.*

PEMMICAN

5 cups pounded jerky
4 cups crushed serviceberries
1/4 cup melted marrow, tallow, or lard
sugar or honey (optional) to taste

Mix first two ingredients well. Gradually add hot fat until mixture is moist enough to stick together. Form into small balls (about golf-ball size). May add honey or sugar to sweeten.

A Crow recipe, Theo Hugs, Ft. Smith, Montana

CANNED JUNEBERRIES

Wash and pick over berries. Pack into sterilized pint jars to within 1 inch of top. Add 1 tablespoon lemon juice.

Fill to within 1/2 inch from top of jar with hot syrup. (1 cup of water to 2 cups sugar. Boil until sugar dissolves. Do not boil down but keep hot as you pack berries). Screw rings on firmly. Process in hot water bath for 15 minutes.

Barb Griffith, Helena, Montana

SERVICEBERRY COFFEE CAKE

1/2-3/4 recipe yeast-type coffee cake dough
1 cup serviceberries
1 cup milk mixed with 2 tablespoons dry milk
2 eggs
1/2 cup brown sugar or 3/8 cup honey
dash of nutmeg
1/2 cup fine bread crumbs
1/2 cup chopped walnuts or sunflower seeds
3/8 cup brown sugar

Make coffee cake dough as usual. Roll into a circle to fit bottom and side of 9'' pie pan. Prick dough with fork to prevent excessive rising. Sprinkle serviceberries on dough. Beat eggs, milk, and nutmeg and pour over berries. Mix last 3 ingredients and sprinkle over custard mixture. Bake at 375 degrees for 25 minutes.

Theo Hugs, Ft. Smith, Montana

Wild Grape

Wild grape was not on my berry-picking list in my younger years. Though they grew in Michigan, my mother didn't want to be bothered with the pea-sized grapes when she could have Concord grapes straight from our own vines.

I had always thought that wild grapes were confined to the United States east of the Mississippi. It wasn't until 1987 that I discovered a wealth of them growing in the canyons of the Big Horn Mountains. That fall I managed to pick enough wild grapes to put up three batches of wild grape jelly.

My sons loved it and went through the batches in three months. From then on, they clamored for grape jelly. Whenever I would call and ask if they needed anything, they would respond "grape jelly." Mind you, that's a pretty amazing order when you consider how teenagers demand all sorts of material goods.

Unfortunately, 1988 was a devastating drought year. I did not find enough grapes to bother to pick. My sons were bummed—especially since I wasn't able to come up with enough chokecherry jelly and wild plum jam to keep them happy.

Fortunately, 1989 was a fantastic year for berries. I managed to pick enough grapes to do two batches of jelly. Then frost hit and eliminated the grapes I had intended to pick.

One day in early October, I was guiding a couple, Rick and Tina. A fantastic mayfly hatch was occurring, causing trout to rise all over the river. I floated about a half-mile from my launching point and rowed to shore at a high bank. I gave Rick the correct fly and pointed out several rising fish that were in easy casting range. He set out.

I turned and asked Tina if she wanted to fish. She said no. I was in a quandary until I spotted the vines draping the bank with deep blue patches among the vines. I asked her if she liked to pick berries. She said she'd love to, but where were they? I pointed to the vines, grabbed my berry buckets from the boat hatch and we embarked.

It took about ten minutes for us each to pick a gallon of grape clusters. I went back to the boat, grabbed a plastic garbage bag and emptied the buckets into it. In a little over a half hour, we had picked another six gallons of grapes.

About that time, Rick came trooping back. He had caught a half dozen trout in the sixteen to nineteen-inch range and was proud as could be. Tina showed him her trophy: a plastic bag full of wild grapes!

Tina had told me that she grew up in Illinois where she used to pick berries with her mother. Picking the wild grapes with me had given her a wonderful nostalgic trip.

That evening, I pulled about two gallons of the grapes off the stems, steamed them up, rendered the juice, and made a batch of jelly. I presented Tina with two twelve-ounce jars the next morning as a token of my appreciation for her help. Thanks to her, I knew that my sons would not run out of grape jelly during the winter!

Wild grapes occur on the east slope of the Rocky Mountains all the way east to the Atlantic. There is also a canyon grape species that occurs from southern Colorado and Utah into Mexico. The one species found in Wyoming, Montana, and Colorado, the fox grape, *Vitis riparia*, grows in the foothills and canyons and along streams on the plains. It has a shredded, brown bark. The vines climb quite readily. If it is growing among trees or tall shrubs, expect to use a ladder to pick the grapes. Grape vines have tendrils at the nodes which enable them to climb.

The leaves are light green in color, deeply toothed, and shallow-lobed. Overall, they have a shape and size much like that of a sugar maple leaf.

Grapes have fairly small, inconspicuous, greenish-yellow blossoms which bloom in May.

Grapes ripen in September but will linger on the vines until a frost kills them. The grapes are not much bigger than peas, have a deep blue color with a powdery blush, and occur in clusters of ten to twenty or so.

I have found it is easiest to pick the entire grape cluster and wait until I get home before I stem the grapes. Incidentally, grapes stain your hands indelibly. If you want to get the stains off your hands, counter, or porcelain sink, use chlorine bleach.

You can simmer the grapes with two cups of water or so to each gallon of grapes for about twenty minutes. I usually run the cooked grapes through my Foley food mill a cup at a time to remove the juice and pieces of the skin. I then can either make jelly from the pulpy juice or freeze it in gallon milk containers until I can get around to it.

Your chances of picking enough grapes to make jelly are good—if you know where the grapes are. A little preseason scouting will help guarantee your success. If you have scouted the wild grapes, your chances are nine out of ten that you will get two gallons. Without preseason scouting, I would say that your chances are fifty-fifty. 🐛

Recipes

GRAPE CONSERVE

4 pounds wild grapes
2 pounds sugar
1/4 teaspoon salt
1 cup seedless raisins
1 orange
1 cup finely chopped nuts

Wash and drain the grapes, remove them from the stems, slip off the skins, and keep them separate. Cook the grape pulp for about 10 minutes, or until the seeds show. Press the grapes through a sieve to remove the seeds. To the pulp add sugar, salt, raisins, and orange which has been chopped fine, rind and all, and had the seeds removed. Cook rapidly until the mixture begins to thicken, stirring frequently to prevent sticking or scorching. Add the grape skins, cook for 10 minutes longer or until the conserve is thick. Stir in the chopped nuts, and pour at once into sterilized jelly glasses. When cold, cover with melted paraffin, and store in a cool, dry place. (The boiling water bath method is recommended for safe food storage).

Laale Cina, Cody, Wyoming

WILD GRAPE JELLY

5 cups grape juice
7 cups sugar
4 ounces pectin

Gather wild grapes when fully ripe, around the first of September. Rinse and discard large stems so that grapes will settle compactly in stewing pot. Cover with water and simmer about 15 minutes. Drain through cloth to obtain juice.

In a 6- to 8-quart pot or kettle, add 5 cups juice to pectin, bring quickly to a boil while stirring constantly. Add sugar and bring to a full rolling boil. It is important to stir constantly to prevent scorching. Boil hard for one minute. Remove from heat. Skim off foam with metal spoon. Pour at once into sterilized jelly glasses. If sealing with wax, leave 1/2-inch at top and cover jelly at once with 1/8-inch paraffin. Two-piece lids may be used, leaving 1/8-inch space at top. Place lid on jar, screw band on; invert jar. When all are sealed, stand upright; cool. Store in cool place.

Laale Cina, Cody, Wyoming

WILD GRAPE BUTTER

6 quarts stemmed and washed grapes
water (enough to cover)
4 quarts apples
4 cups sugar

Cover the washed grapes with water and simmer for 20 minutes. Strain off juice and make into jelly. Put the grape pulp into a cheesecloth bag. Return to the kettle in the bag. (The bag keeps the grape seeds out of the apples, but gives a grape flavor to the butter). Add apples which have been quartered, but not peeled.

Cover with water. Bring to a boil, then simmer 20 minutes. Drain. Juice can be used for Grape/Apple Jelly. Put apples through sieve and measure 5 cups. Place in kettle, add sugar and heat to boiling, stirring constantly. Cook to desired consistency. Pour into hot jars. Process in boiling water bath to ensure a good seal.

Margaret and Charles Butterfield,
Preserving Wyoming's Wild Berries and Fruits,
University of Wyoming Agricultural Extension Service, 1981.

GRAPE/APPLE JELLY

5 cups grape-apple juice (see previous recipe)
7 cups sugar
1 package powdered pectin

Measure the juice and add it to a large kettle or pot (at least 4 quarts). Add powdered pectin and stir to dissolve, then bring to a quick, hard boil over high heat, stirring occasionally. Add premeasured sugar all at once. Bring to a full rolling boil (a boil that cannot be stirred down). Boil hard for one minute, stirring constantly.

Pour into sterilized jars. Fill to within 1/8-inch of the top, wipe rim clean, place hot metal lid on jar, and screw metal band down firmly. Process in boiling water bath for 5-10 minutes.

Margaret and Charles Butterfield,
Preserving Wyoming's Wild Berries and Fruits,
University of Wyoming Agricultural Extension Service, 1981.

GRAPE JUICE

Wash and stem fresh, firm-ripe grapes. Put 1 cup grapes into a hot quart jar. Add 1/2 to 1 cup sugar. Fill jar with boiling water, leaving 1/4-inch head space. Adjust cap. Process quarts 10 minutes in boiling water bath.

Margaret and Charles Butterfield,
Preserving Wyoming's Wild Berries and Fruits,
University of Wyoming Agricultural Extension Service, 1981.

SPICED WILD GRAPE JELLY

8 quarts wild grapes
1 quart vinegar
1/4 cup whole cloves
1/4 cup stick cinnamon
12 cups sugar

Combine the first 4 ingredients in a large kettle. Heat slowly to the boiling point. Cook until the grapes are soft. Strain off juice, then boil it in a kettle for 20 minutes. Add heated sugar (put sugar in oven and heat), boil for 5 minutes. Pour into sterilized glasses and seal. Place in a boiling water bath and process for ten minutes. Yield: Twelve 8-ounce jars.

Betty Bindl, Trout Creek, Montana

HONEY WILD GRAPE JELLY

2 1/4 cups wild grape juice
1 package MCP pectin
3 1/2 cups honey
1/4 teaspoon margarine, butter, or cooking oil

1. Measure honey in bowl to be added later.

2. Measure grape juice into 6- or 8-quart saucepan or kettle. If a little short of juice, add water. If short more than one cup juice, add another type of fruit juice.

3. Add the package of MCP pectin to measured fruit juice. Stir thoroughly to dissolve, scraping sides of pan to make sure all the pectin dissolves. (This takes a few minutes.) Place mixture over high heat. Bring to a boil, stirring constantly to prevent scorching.

4. Add the premeasured honey and mix well. Continue stirring and bring to a full rolling boil (a boil that cannot be stirred down). Add the 1/4 teaspoon of margarine, butter, or cooking oil and continue stirring. Boil hard exactly 2 minutes.

5. Remove mixture from heat. Skim foam and pour into glasses. Seal, place in a boiling water bath for 10 minutes. Yield: five 8-ounce jars.

MCP Foods,
Make a Honey of a Jam with Recipes from MCP Foods,
Anaheim, California.

Wild Plum

Wild plums are one of the West's greatest treasures. Plums provide buckets full of delectable fruit nearly every fall. I like wild plums because they are good-sized—I can pick a berry bucket full in short time. They have a zesty taste and make an unequaled jam.

I really don't feel like I have experienced autumn without at least one good session of plum picking. There's scarcely a more enjoyable activity than walking about the rolling plains and foothills on a sunny day searching for clumps of wild plum that flourish in the draws, canyons, and floodplains. Picking plums seems to satisfy my hoarding instincts, soothe my soul, and give me an artistic feeling. They are such a pretty fruit that a bucketful of them looks like an oil painting by some European master artist.

Plums are one species I watch from spring through summer so that I will be able to find enough when they are ripe. Wild plums are one of the first flowering shrubs to blossom. It seems that they always have a profusion of white blossoms in late April and early May. Unfortunately, the danger of frost hasn't passed when they blossom, so many times the wild plum crop is wiped out by a late cold snap.

The list of maladies grows in the summer. Insects can invade the green plums and cause them to swell up, turn black, and fall off. Drought can cause the plum bushes to drop many of the green fruits.

What amazes me is that there are any at all, with all that can go wrong with a plum crop. Yet, with the proper advance scouting, I usually can find several thickets of loaded plum bushes. Some of the bushes are so bountiful that the branches are bent to the ground. I always consider it my duty to help those laden bushes become upright again.

Wild plums have a nice aroma to them. I like to pick them when they are still a tad green, bring them home, and let them ripen in a closet for a few days before I put them up into jam.

When I open the closet, the plum fragrance wafts gently to my nose. It is a delicate, sweet fragrance that means good eating ahead.

Wild plum has a scientific name of *Prunus americana*. Chokecherry, pin cherry, black cherry, and beach plum belong to the same genus and, of course, to the same family, *Rosaceae*.

Plums, black currants, gooseberries, Oregon grapes, and serviceberries are the earliest blooming shrubs. You can find plums blossoming in late April to early May. Since the areas that plums inhabit—drainage areas such as draws, ravines, canyons, and stream floodplains of the plains and foothills—differ little in altitude, there is but two weeks or so divergence in their flowering throughout their range. Plums usually ripen late August through September. If a frost doesn't knock them off, they will stay on the bush into October. Ripe plums vary in size from that of a marble to that of a cherry tomato—it all depends on how fertile the soil is where the bush is growing and the amount of moisture it receives. The plums are round to slightly oval in shape. A ripe plum can vary in color from deep pink with a blush, to pale yellow.

Wild plum bushes vary in height from three to ten feet. Some of the books I have read state that wild plum reaches a height of five meters (16. 5 feet), but I have never seen a wild plum bush that tall. Most of the time the bushes are clumped together in thickets that range from wading pool-to motel swimming pool-sized.

The older branches are glossy gray with thorns or spines at the tips; the younger branches are a rich brown color. The leaf is two to four inches long, lanceolate (spear-shaped), with a serrate margin.

Wild plum is an east slope plant. In fact, you can find wild plums growing from Montana-Wyoming-Colorado to the Atlantic Coast.

Wild plums can be the most abundant wild fruit in the area. You can pick a larger quantity of plums in a shorter time than any other berry or fruit mentioned in this book. In good years, you should be able to pick five gallons of wild plums in two hours. Your chances of doing that are about nine out of ten in most years. You can increase your success rate by preseason scouting.

Wild plums are easy to freeze. Simply wash them, let them drain dry, and then place in gallon freezer bags. They'll keep this way for six months or so—plenty of time for you to make them into jam, jelly, cobbler, pie, or fruit leather. The thawed wild plums will be much easier to pit: simply squeeze them and the pit will come flying out.

Another way of pitting plums is to use a cherry pitter. This machine can make quick work of pitting plums. You can purchase one from a cookware specialty store, or many seed companies market them in their catalogs.

I usually pit fresh plums for making jam by putting the plums in a kettle, adding two cups of water, bringing them to a boil, then simmering them for fifteen minutes or until they start to burst. I let them cool, then just squeeze the plum to get the pit out. 🐾

Recipes

WILD PLUM JELLY AND BUTTER

Wash and pick stems of wild plums. Put in water just to cover or less if plums are very juicy; boil until soft, dip out juice with a china cup. Then strain rest through loosely woven cloth. Do not squeeze them. Take pound for pound of juice and sugar, or pint for pint, and boil for 8 minutes. Jelly will be nicer if one measure, or measure and a half, is made at one time; if more, boil longer. It can be tested by dropping on a saucer and placing on ice or cool place. If it does not spread, but remains rounded, it is finished. Stir while cooking and skim before pouring into jars.

For plum butter, take the plums that are left and press through a sieve. Take pint for pint of plums and cooked dried apples. Take pint for pint of pulp and sugar and boil about a half hour, testing as above for jelly, as this will also jell. Cook and stir as above, skimming before sealing. Seal with wax and cover. Store in a cool place.

Laale Cina, Cody, Wyoming

WILD PLUM PUDDING CAKE

2 1/2 cups flour
1 cup sugar
3 teaspoons baking powder
1 teaspoon salt
1 cup milk
1/4 cup shortening

Combine the dry ingredients then mix in the milk and shortening. Spread in a 9" x 13" baking pan.

Drain (save juice) and pit 2 quarts of canned wild plums. Sprinkle plums on top of batter.

Sauce:
4 cups juice (add hot water to get total)
1 1/2 cups sugar
red food coloring
1 teaspoon cinnamon
4 tablespoons margarine

Bring sauce to a boil and pour over plums. Bake at 350 degrees for 30 minutes. Sauce will be on the bottom and crust on top when done.

Ruby Montgomery, Sheridan, Wyoming

RASPBERRY-PLUM JELLY

4 cups raspberries, fresh or frozen
2 cups pitted, cut up wild plums
1 medium lemon
sugar

Combine raspberries and cut up plums. Quarter the lemon, then slice very thin, discarding seeds. Add to the raspberries and plums. Place all in a heavy kettle and crush fruit with the base of a heavy glass tumbler. Add 1/2 cup water and cover kettle. Simmer until fruit is very tender, then drain in a jelly bag. Measure juice and add 3/4 as much sugar. Place over high heat and bring to a full, rolling boil, stirring frequently. Cook rapidly until jelly sheets (220 F.). Remove from heat and skim. Seal in hot sterilized glasses. About 6 small glasses.

Bob Giurgevich, Sheridan, Wyoming

WILD PLUM WINE

1 gallon wild plums
1 gallon water
8 pounds sugar
small box raisins
⅓ package cake yeast

Mix together in anything but a metal bucket—work at least once a day for approximately 3 weeks. Strain and bottle.

Dean Davis, Sheridan, Wyoming

PIONEER PLUM JAM

For every cup of plum pulp (with skins) add 3/4 cup sugar. Cook over low heat until of desirable consistency for spreading. Stir often to prevent scorching. The mixture thickens when cool.

Fill hot jars to within 1/4 inch of top with hot mixture. Wipe rim clean. Adjust lids and process in boiling water bath for 5 minutes.

Margaret and Charles Butterfield,
Preserving Wyoming's Wild Berries and Fruits,
University of Wyoming Agricultural Extension Service, 1981.

CANNING PLUMS

Whole plums can be canned and used as winter fruit served plain or with cream. Wash plums and discard those which are wormy or spoiled. Heat to boiling in syrup (2 cups sugar and 4 cups water) or in water. Pack hot fruit to 1/2 inch of top of jars. Cover with boiling liquid, leaving 1/2 inch space at top. Adjust jar lids. Process in boiling water bath 20 minutes for pint jars; 25 minutes for quart jars.

Margaret and Charles Butterfield,
Preserving Wyoming's Wild Berries and Fruits,
University of Wyoming Agricultural Extension Service, 1981.

PLUM PIE

4 cups sliced, pitted plums
1/2 cup sugar
1/4 cup flour
1/4 teaspoon salt
1/4 teaspoon cinnamon
1 tablespoon lemon juice
1 pie crust (see serviceberry recipes)

Cut plums in quarters. Combine with sugar, flour, salt, and cinnamon. Turn into pie shell and sprinkle with lemon juice. Add Spicy Topping—

Spicy Topping:
1/2 cup flour
1/2 cup sugar
1/4 teaspoon cinnamon
1/4 teaspoon nutmeg
1/4 cup butter or margarine

Cut in butter until mixture resembles coarse crumbs. Sprinkle over plums, mounding topping up in center of pie. Place in brown paper bag. Fold bag over twice and fasten with paper clips. Place on cookie sheet. Bake at 425 degrees for 1 hour.

Judy Faurot, Sheridan, Wyoming

WILD PLUM SAUCE

Pick wild plums when they have a nice color and are still hard. (Fully ripe plums will have hard inedible skins when cooked.) Cold pack plums, pour on water to overflowing, and boil jars without lids for 20 minutes. Pour off juice and keep the fruit in the jars. Fill to within 2 inches of the top with a medium syrup (1 cup sugar to 2 cups water). Seal and process for 20 minutes in a boiling water bath.

Ellen Ricki, Lewistown, Montana.
From the Billings Gazette Cookbook, *September 25, 1983.*

WILD PLUM PIE

2 cups pitted plums, cut finely
1 cup sugar
2 cups shredded apples
1/2 teaspoon almond flavoring
2 tablespoons flour or cornstarch
1 cup cream
1 pie crust (see serviceberry recipes)

Mix all ingredients and let stand for 10 minutes. Pour in unbaked pie crust and bake 1 hour at 375 degrees or until done.

Alida Herigstad, Hodges, Montana.
From the Billings Gazette Cookbook, *September 25, 1983.*

WILD PLUM FRUIT ROLLS (LEATHER)

4 cups wild plum puree
1 package MCP pectin
1 cup sugar

1. Use fully ripe or slightly overripe plums. Wash and cut away any bruised or spoiled portions. Pit.

2. Puree plums in blender or food processor.

3. Stir the MCP pectin into puree. Mix well. Add sugar and stir until dissolved.

4. Spray or coat cookie sheet or dehydrator shelf with vegetable oil. Spread 1 cup puree in border pattern. Smooth puree with rubber spatula or tilt cookie sheet to evenly spread puree. Refrigerate unused puree.

5. For conventional oven: Set temperature control at lowest temperature or 150 degrees F. Two cookie sheets may be placed in the oven at the same time. Rotate trays after 1 1/2 hours. Drying will take approximately 2 1/2 to 3 hours.

For dehydrator:
Set temperature control at 140 degrees F. and dry for 6-10 hours.

For sun-drying:
Use same techniques as would be used for drying fruits.

6. Rolls are done when slightly sticky to the touch, but dry and pliable.

7. Remove rolls from tray while still warm and either roll each one in one piece or cut them into 4- or 6-inch squares. Roll in plastic wrap. Rolls may be stored up to 4 months without refrigeration. For longer storage, place in refrigerator up to 1 year or in freezer.

> *Special Recipes from MCP Foods*
> *P.O. Box 3633, Anaheim, CA 92803.*

WILD PLUM PRESERVES

Wash plums, place in kettle, and cover with water. Cook until tender. Cool. Pit by hand. (Some folks run through a colander, but that's lots of work plus I like the skins wholer!) Save the juice and leave it with the pitted plums.

Mix by volume 50/50 with sugar. Cook over high heat until desired consistency.

Pour into sterilized jars, cover with two-piece lids, and process in boiling hot water bath—10 minutes for pints; 20 minutes for quarts.

These preserves are great with fresh cream.

> *Mary Moravek, Sheridan, Wyoming*

PLUM-CRAB APPLE CONSERVE

2 cups diced crab apples (cored)
2/3 cup crushed pineapple (drained)
1 quart pitted wild plums
5 1/2 cups sugar
1/3 cup shredded almonds

Place crab apples, drained pineapple, and plums in a kettle. Mix together, place over high heat, and bring to a boil. Add sugar, stir constantly. Cook to desired consistency. Remove from heat, stir, and skim. Add almonds. Pour into hot, sterilized jars, put on two-piece lids, and process in hot boiling water bath. Makes 7 half-pints.

Linda Linn, Rock Springs, Wyoming

PLUM HONEY

2 gallons plums, pitted. Scald with 1 teaspoon soda. Set until cool. Pour off juice and use for jelly. Rub plums through colander and mix with sugar (2 cups sugar to 1 cup of pulp).

Leave in kettle and beat several times a day until clear (may take several days). Put in sterilized pint jars, seal, and process in boiling water bath for 20 minutes.

Evelyn Hejde, Aladdin, Wyoming

FREEZER PLUM JAM

2 1/4 cups prepared pulp (2 pounds wild plums)
1 package MCP pectin
4 cups honey

To Prepare Plums:
Wash. Pit, slice, and finely grind.

To Make Jam:
1. Measure prepared pulp into large saucepan.

2. Slowly stir in 1 package of MCP pectin and mix thoroughly. Set aside for 30 minutes, stirring every 5 minutes to dissolve pectin completely.

3. Add 4 cups honey. Mix well.

4. After honey is thoroughly dissolved, pour jam into containers and seal. Store freezer jam in freezer. Refrigerate after opening.

Yield: Six 8-ounce jars.

MCP Foods,
Make a Honey of a Jam with Recipes from MCP Foods
Anaheim, California.

Wild Rose

Some plants make living a joy. Wild rose is one of them. Have you ever hiked through an area where the wild rose was blossoming? The fragrance pleases the soul. When Shakespeare made the statement "What's in a name? That which we call a rose by any other name would smell as sweet," he knew the sweet fragrance of wild rose.

I have seen riverbanks, the bottomlands, and hillsides along creeks covered with wild rose blossoms. The hundreds of delicate pink blossoms made for massive pastoral paintings of the countryside.

Still, with all the times I have witnessed wild roses blooming, I never thought about harvesting them or their fruits (called hips). I knew that rose hips had the highest concentration of natural vitamin C of all plants, and that teas were concocted out of the hips, but I didn't know that jelly was made from the hips. It wasn't until I met a down-home, country lady, Mary Moravek, in 1980 that I realized that rose hips did make great jelly, and it didn't take an inordinate amount of work to do it.

Mary and another lady, Linda Linn, spent one day picking rose hips and making them into jelly. They managed to pick enough rose hips to make two batches despite all the thorns they had to wade through. They wore heavy pants and leather gloves to overcome the thorns. They also kept an eye out for rattlesnakes, since the ladies knew that rattlers liked rosebushes for cover.

While Mary and Linda picked the two-acre patch of wild rose, a doe mule deer and her fawn watched from the center of the patch. The deer seemed to know the berry pickers wouldn't harm them, but they couldn't figure out why humans would tromp around through that thorny patch.

Mary and Linda finished out the day by cleaning the rose hips, and preparing rose hip jelly. It was pleasantly sweet, and had a delicate flavor.

Mary always had recipes for every imaginable food. She also had sage advice for her friends. She had a heart twice the size of Sheridan County, yet she wouldn't hesitate to jump someone if she thought that they were full of horsefeathers. She was a person I felt proud to know and to be listed as her friend. She passed away in 1983.

Maybe the reason I identify Mary with wild rose is that she had beauty, and, like the rose, she was tough and resilient.

Wild rose has plenty of thorns. It grows in areas where most shrubs won't—it needs just a little bit more moisture than a bunch grass to survive.

Several species of wild rose (genus *Rosa*) grow in the West, but they all have the same characteristics: five pink petals, the same type of fruit (hip), and thorns or prickles. Some species may reach four or five feet in height, others are ground hugging and don't exceed a foot. All have alternate, pinnately compound leaves—the leaflets have serrate margins and vary in number from five to nine leaflets.

Wild rose blossoms from June through July depending on elevation. It ripens late August through September. The hips turn to a lustrous red or orange when ripe. The hips can be either globular or elliptical in shape. Most species will have five rather stringy sepals attached at the tip.

Wild rose blossoms from June through July depending on elevation. It ripens late August through September. The hips turn to a lustrous red or orange when ripe. The hips can be either globular or elliptical in shape. Most species will have five rather stringy sepals attached at the tip.

Wild rose is rather widespread in the West. You can find it growing on the plains, along ravines and draws, along watercourses, and into the foothills and mountains. If there are some minor seeps, you will find it growing in the deserts like Wyoming's Big Horn Basin.

Rose hips can be dried for later use in tea or wine, but if you are making jelly, I'd say it would be best if you rendered the juice within a week or so after picking the hips. The hips will keep in the refrigerator for two weeks.

Your chance of finding a quart or two of rose hips is excellent. I would rate your chances at ninety-five out of a hundred in most years. 🍒

Recipes

ROSE HIP JAM

1. Cut hips in half—remove seeds
2. Grind 5 cups cleaned hips.
3. Boil with just enough water to cover hips
4. Rub through a sieve
5. Add juice of 1 lemon
6. Add 3 1/2 cups sugar
7. Cook until thick
8. Pour into sterilized jars and seal.

Mary Moravek, Sheridan, Wyoming

ROSE HIP JELLY

6 cups rose hips
2 cups water
4 cups sugar
1/2 bottle liquid pectin

Snip the bud ends of the rose hips. Coarsely grind or mince them. Place the pulp in a 2- or 3-quart saucepan and add water. Place over low heat and simmer 1 hour or until the pulp is very tender. Strain through a jelly bag or cheesecloth, pressing down on the fruit with a wooden spoon. Clean the pot and measure into it 3 cups of juice and 4 cups sugar. Bring to a boil, stirring constantly. Add the pectin. Boil for 4 minutes, stirring all the time. Remove from heat and quickly skim foam from the surface. Pour into sterilized jars and seal with paraffin.

Bob Giurgevich, Sheridan, Wyoming

ROSE HIP SYRUP

3 pounds rose hips (ripe)
1 cup honey

Wash hips, remove stems and ends. Use stainless steel or enamel saucepan. Simmer 15 minutes or till tender, mash with wooden spoon. Simmer another 8 minutes. Pour into several layers cheesecloth, allow to drip overnight into ceramic bowl. Squeeze out leftovers. Return juice to saucepan, add honey, and blend well. Bring to boil; boil 1 minute. Pour into jars and seal.

Adrienne Crowhurst,
The Weed Cookbook,
Submitted by Earl Jensen, Greybull, Wyoming

ROSE PETAL JAM

1 cup rose petals
3/4 cup water
juice of 1 lemon
2 1/2 cups sugar
1 package pectin

Pack 1 cup petals firmly in blender. Add 3/4 cup water and juice of 1 lemon, blend till smooth. Slowly add 2 1/2 cups sugar until dissolved. Stir 1 package pectin in 3/4 cup water. Bring to boil; boil 1 minute. Pour hot mix into blender and blend 1 minute. Pour into jars and seal.

Earl Jensen, Greybull, Wyoming

ROSE HIP AND RHUBARB JAM

Use slightly underripe rose hips. Cut in half and remove seeds with tip of knife.

Combine:
1 cup rose hips
1 cup water
4 cups diced rhubarb
1/4 teaspoon salt

Boil rapidly 1 minute and add:
2 cups sugar
1 tablespoon grated lemon rind

Boil rapidly 1 minute. Seal in sterilized jars.

Mrs. Pauline Deem, Plentywood, Montana

DRIED ROSE HIPS

Dried rose hips can be used to sprinkle over desserts or cereals, added to batters for baked goods, or combined with tea or fruit juices for a hot or cold beverage. Cut rose hips in two and remove seeds with the point of a knife. Dry as quickly as possible in a slightly warm oven or food dehydrator. Crumble dried hips and use as desired.

Mrs. Pauline Deem, Plentywood, Montana

CANDIED ROSE HIPS

1 1/2 cups rose hips
1/2 cup sugar
1/4 cup water

Remove seeds from rose hips. Boil 10 minutes in the sugar-water syrup. Lift fruit from syrup with a skimmer and drain on waxed paper. Dust with sugar and dry slowly in the sun or very slow oven, adding more sugar if the fruit seems sticky. Store between sheets of waxed paper in a closely covered, metal container until used.

Uses for candied rose hips:
In your favorite cookie recipe in place of, or in addition to, nuts or chopped fruit (oatmeal cookies, fruit squares, or filled sugar cookies); in puddings with added grated lemon rind; or in place of nuts or fruits.

Margaret and Charles Butterfield,
Preserving Wyoming's Wild Berries and Fruits,
University of Wyoming Agricultural Extension Service, 1981.

ROSE HIP TEA

Grind approximately 3-4 cups of rose hips. Boil in 2-3 cups of water for 20 minutes. Strain the liquid to remove the pulp. It's delicious either hot or cold.

Jim Bayne, Bozeman, Montana,
From Savoring the Wild,
A collection of favorite recipes from the employees of the Montana Department of Fish, Wildlife, and Parks. Falcon Press, Helena, Montana, 1989.

ROSE HIP SYRUP AND FRUIT BUTTER

Snip bud end from hips. Cover fruit with water and boil until soft. Cool. Grind in blender. Strain juice. For every 2 cups juice, add 1 cup sugar. Boil until thick. Bottle or seal in jars.

Press pulp through a sieve. For every 2 cups pulp, add 1 cup sugar. Add cinnamon, cloves, allspice to taste. Heat to dissolve sugar. Cook, stirring constantly to prevent sticking. Pack in jars and seal. Boil in water bath 10 minutes.

Charlotte Heron, Missoula, Montana

ROSE HIP CANDY

Gather rose hips, grind into a paste, mix with tallow (butter can be substituted), and add sugar to sweeten. Shape into little balls, put a stick into the balls, and roast them over hot coals and enjoy them as a treat on your camping trips.

A Crow recipe,
Theo Hugs, Ft. Smith, Montana

ROSE HIP-CRANBERRY JELLY

1 pint rose petals
2 cups cranberries
1 cup water
4 cups sugar

Fill a pint jar completely with freshly picked wild rose petals. Cover the blossoms with boiling water and cover. Keep out of bright sunshine. Allow the infusion to sit overnight for 24 hours. This will leach the color and flavor from the blossoms. The next day strain the infusion, removing the petals. Clean and remove stems from 2 cups of cranberries and boil in 1 cup of water for about 20 minutes or until juice runs freely. Strain through a sieve.

Place the rose petal infusion and the cranberry juice in a saucepan, bring to a boil, add the sugar, and boil until the sugar has completely dissolved. Remove from the heat, skim off the foam, pour into hot, sterile, jelly jars, seal, and process in a boiling water bath. You may adjust the amount of sugar to suit your taste.

Cel Hope, Sheridan, Wyoming

Wild Raspberry

When raspberries were created, the perfect berry came to this earth. Well, maybe that is a tad bit of exaggeration, but raspberries are heavenly. About their only sin is that there aren't enough of them every year to satisfy the humans craving them.

I have had some wonderful raspberry-picking trips all over Wyoming. Some patches in Jackson Hole contain stupendous amounts of berries. There are some great patches in the Big Horn Mountains, too, but I'm not telling where. You see, many folks think that a good wild raspberry patch is to be well guarded—you don't tell someone outside the immediate family where you go for raspberries.

A fellow that used to teach various wild plant courses for Sheridan College, Al Dumont, told me about some of his berry hot spots. Mind you, he just told me about them, he never told me any *exact* locations. He contended that one island in a river in the Big Horn Mountains had so many wild raspberries on it that he named the place "Raspberry Island." I have pored over my topographic maps of the Big Horns and have yet to find any spot named such, but I'll continue looking. Dumont contends that the place is so good that it's worth the search—even if I have to walk every stream in the Big Horns.

One of the most productive raspberry patches I ever saw was along a timbercut. The ripe raspberries stretched the length of the clearcut, coloring the area red. My friend, Dot, and I picked two quarts so easily that it seemed to make up for the many times that I have worked hours for a pint. About the only problem we had was worry about other berry pickers—there was ample bear sign around the area and we certainly didn't want a confrontation with one of them!

Wild raspberries are coveted by man and beast. Bears, other mammals, songbirds, and game birds relish wild raspberries, so it is important to have several patches in mind before you go after wild raspberries. It is almost certain that someone or some critter will have beat you to the patch. When you do find a good patch, you will think that you have discovered a treasure.

Wild raspberries love sunlight and don't compete well with other shrubs or trees. They grow from the foothills to the alpine zone on disturbed areas such as timber cuts, talus slopes, burns, and stream floodplains.

Raspberries are biennials, that is, each cane (stem) lives two years. The rootstock sends up shoots each year which turn into canes. The second year the canes branch out a bit, blossom, and bear fruit. Shortly after the berries ripen, the canes die.

Raspberry canes have small thorns or prickles. Some canes have few prickles, others are quite "hairy" with prickles. Most wild raspberries reach a height of fifteen inches to three feet. I have read where some botanists claim that wild red raspberry reaches heights of over six feet, but I have never seen any wild raspberries that tall.

Raspberries have white-petaled inflorescences blooming from June at lower altitudes into July at higher altitudes. The half-hemisphere shaped berries, about half the size of domestic raspberries, ripen about a month and a half later. If you have ever seen domestic raspberries, you'll have no trouble identifying wild raspberries, for they look much the same.

Wild raspberry leaves are pinnately compound (that is, the leaflets are arranged along the main axis of the leaf) with either three or five leaflets. The undersides of the leaflets are whitish.

Raspberries belong to the genus *Rubus*. The most common species in the Rocky Mountain West is *R. idaeus*.

Your chances of getting enough raspberries for any major product like jam—which requires two quarts—are slim, probably no more than one chance in five. In normal years, you stand one chance in two of getting a pint.

Raspberries do store well. If you want to store a couple of pints in the freezer, wash and drain them dry on several thicknesses of paper towels, place in plastic freezer bags, and toss them in the freezer. They will be mushy when they thaw, but they will still taste great. 🦌

Recipes

DUTCH OVEN DELIGHT

This camping recipe is Earl Jensen's own creation. Any wild berries can be used, a mixture of different berries or just one kind. Put as many berries as you want in the bottom of a greased dutch oven. Barely cover with water, add sugar to taste, and dissolve. Once dissolved add and stir Bisquick until all water is absorbed or nearly so. Cover oven, put on hot coals, and put coals on lid and around sides—not too many along sides. A little experimenting and you will come up with a tasty dessert. Don't use too hot a fire or center will be raw; slow cooking is better.

Earl Jensen, Greybull, Wyoming

RED RASPBERRY JAM CAKE

1 cup brown sugar
3/4 cup butter (creamed)
3 eggs
1 1/2 cups flour
3 tablespoons sour cream
1 teaspoon nutmeg
1 teaspoon cinnamon
1 teaspoon allspice
1 teaspoon baking soda
1 cup wild red raspberry jam

Mix ingredients in order listed. Bake in 3 layers until done. (350 degrees for about 30 minutes). Frost with white frosting.

Judy Faurot, Sheridan, Wyoming

RASPBERRY-CURRANT SHERBET

2 pints fresh raspberries
1 1/4 cups currant jelly
2 cups half and half or light cream
1/2 cup creme de cassis

Puree raspberries in blender. Place puree and jelly in a 2-quart pan. Heat on low till jelly melts, stirring often. Cool to lukewarm. Stir in half and half and creme de cassis. Chill 1 hour. Churn freeze. Place in plastic freezer container. Let ripen 3 hours in freezer before serving.

The Missoulian, 1985,
Submitted by Charlotte Heron, Missoula, Montana

VERY BERRY SORBET*

1 envelope unflavored gelatin
1/2 cup sugar
1 1/2 cups water
2 cups pureed raspberries
1/2 cup creme de cassis or cranberry juice
2 tablespoons lemon juice

In a medium-sized saucepan mix unflavored gelatin with sugar. Blend in water. Let stand 1 minute. Stir over low heat till completely dissolved. Let cool to room temperature. Stir in remaining ingredients. Pour into a 9'' baking pan and freeze for 3 hours or till firm. With electric mixer or food processor, beat till smooth. Return to pan and freeze 2 more hours or till firm. To serve, let thaw at room temperature 15 minutes or until slightly soft. Makes 8 servings.

*Strawberries, blackberries, juneberries, or huckleberries can be substituted.

The Missoulian, *1985,*
Submitted by Charlotte Heron, Missoula, Montana.

RASPBERRY-PLUM BUTTER

1 1/2 pounds pitted plums
1 quart raspberries
1/2 cup water

Bring above ingredients to a boil, reduce heat, and cook till tender, about 10 minutes. Cool. Puree in blender or food processor. Return to kettle and add:

1 1/2 cups sugar
1 tablespoon lemon juice

Cook over low heat till sugar dissolves. Bring to boil and stir constantly till thick and glossy (about 10 minutes). Pour into hot jars. Seal. Process in hot water bath for 15 minutes. Makes 2 pints.

Charlotte Heron, Missoula, Montana

RASPBERRY TWIRLS

2 cups flour (sifted)
1 teaspoon salt
3 ½ teaspoons baking powder
2 tablespoons sugar
2 heaping tablespoons shortening
7/8 cup milk
2 tablespoons butter
1 cup ripe wild raspberries
½ cup sugar

Sift dry ingredients (flour, salt, baking powder, and 2 tablespoons sugar) into a bowl. Cut in shortening. Gradually mix in the milk.

Roll out the dough to a rectangle 1/2 inch thick. Sprinkle the dough with bits of butter. Blanket dough with the raspberries. Scatter 1/2 cup sugar over the berries.

Roll up the berry-covered dough and cut into 1/2 inch slices—you should come up with 10-12 slices.

Butter a large baking pan or casserole. Put the slices in the pan and pour the following sauce over.

1 cup sugar
1 tablespoon flour
dash of salt
1 cup water
1 tablespoon butter
1 teaspoon vanilla

Measure sugar, flour, and salt into a bowl. Mix well. Add water to saucepan. Slowly add dry ingredients to mixture and mix thoroughly. Add butter and bring to a boil. Boil four minutes. Add vanilla and pour over the slices.

Bake at 425 degrees 10 minutes, then at 350 degrees 20 minutes. Serve warm, topped with whipped cream or ice cream.

Dennis Cook, Cody, Wyoming

HONEY-RED RASPBERRY JELLY

2 1/4 cups berry juice (approximately 2 quarts raspberries)
1 package MCP pectin
3 1/2 cups honey
1/4 teaspoon margarine, butter, or cooking oil

To Prepare Juice:
Wash and crush berries thoroughly until reduced to a pulp. Heat to a boil.
Extract juice by placing a colander in a bowl or kettle. Spread cloth or
jelly bag over colander. Pour hot prepared berries into cloth or bag. Fold
cloth to form bag and twist from top. Press with potato masher to extract
juice. Measure 2 1/4 cups juice.

To Make Jelly:
1. Measure honey in bowl to be added later.

2. Measure raspberry juice into 6- or 8-quart saucepan or kettle. If a little
short of juice, add water. If short more than one cup juice, add another
type of fruit juice.

3. Add the package of MCP pectin to measured fruit juice. Stir thoroughly
to dissolve, scraping sides of pan to make sure all the pectin dissolves.
(This takes a few minutes.) Place mixture over high heat. Bring to a boil,
stirring constantly to prevent scorching.

4. Add the premeasured honey and mix well. Continue stirring and bring
to a full rolling boil (a boil that cannot be stirred down). Add the 1/4
teaspoon of margarine, butter, or cooking oil and continue stirring. Boil
hard exactly 2 minutes.

5. Remove mixture from heat, skim foam, and pour into glasses. Seal with
two-piece metal lids and process in boiling water bath for 10 minutes. Yield:
Five 8-oz. jars.

MCP Foods,
Make a Honey of a Jam with Recipes from MCP Foods.
Anaheim, California.

RED RASPBERRY FREEZER JAM

2 1/4 cups red raspberry pulp (approximately 1 quart berries)
1 package MCP pectin
4 cups honey

To prepare fruit: wash, stem, and crush.

To make jam:
1. Measure prepared pulp into large saucepan.

2. Slowly stir in 1 package of MCP pectin and mix thoroughly. Set aside for 30 minutes, stirring every 5 minutes to dissolve pectin completely.

3. Add 4 cups honey. Mix well.

4. After honey is thoroughly dissolved, pour jam into containers and seal. Store freezer jams in freezer. Refrigerate after opening. Yield: Six 8-ounce jars.

MCP Foods,
Make a Honey of a Jam with Recipes from MCP Foods,
Anaheim, California.

RASPBERRY CANDY

1 cup raspberry pulp (about 1 pint red raspberries)
1 package MCP pectin
1/2 teaspoon baking soda (packed, level)
1/2 teaspoon butter or shortening
1/2 cup light corn syrup (Karo)
1 3/4 cups sugar
5 tablespoons lemon juice
1/2-1 cup nuts (optional)

Measure raspberry pulp into large kettle. Stir in MCP pectin. Add baking soda. Stir well to distribute thoroughly and to prevent soda from reacting in spots and darkening juice. Place over heat. Add butter or shortening to reduce foaming. Heat to full boil. Add light corn syrup and sugar. Bring back to full boil, and boil vigorously for exactly 5 minutes, stirring constantly. Remove from heat and add lemon juice. Add nuts.
Pour into 9-inch oiled pie or square pan. The depth should be about 1/2 inch. Allow to harden 24 hours. Cut into squares. Dust pieces with sugar to prevent sticking. Yield: One 9-inch pan.

MCP Foods,
Special Recipes from MCP Foods,
Anaheim, California.

THAT RASPBERRY THING

1 box vanilla wafers (12-16 ounces) crushed
1/2 pound butter (1 cup)
1 pound powdered sugar
4 eggs
4 cups raspberries (fresh) washed and well-drained
1 pint whipping cream

Spread half of the crushed vanilla wafers (about 2 cups) in bottom of 12" x 17" or two 9" x 9" pans.

Beat butter until creamy. Blend in powdered sugar. Add eggs one at a time, beating well after each. Drop butter mixture in small spoonfuls over the crumbs and spread evenly. This is thick so be careful when spreading. Add raspberries in an even layer.

Whip the cream. Add sugar to taste if desired. Spread whipped cream over berries, top with remaining crumbs. Refrigerate at least 12 hours before serving.

Rita Quinn, Portland, Oregon

Wild Strawberry

One of the most memorable weekends I ever spent occurred in 1978. My sons were seven at the time and were fired up to go camping. I got my backpack and the various gear needed for a weekend jaunt, grabbed my fishing gear, picked up the boys, and off we went.

We headed up the Hoback River, past Bondurant and finally turned off on a Forest Service road. After two or three miles, I parked my vehicle, donned the backpack, and we were on our way. I knew the boys weren't going to want to hike more than a couple of miles and would want to fish and play most of the time. The place I had in mind was but a short hike away and along a creek lined with beaver ponds inhabited by small, voracious brook trout that would attack anything the boys threw at them. What I didn't know was that there was a tremendous patch of wild strawberries in the gravelly floodplain of the creek.

The hike only took an hour. The boys soon tired of catching brook trout and wanted other diversions. The strawberry patch gave them that.

They went into the patch with their hiking cups, sat down, and started to fill them. They managed to put one out of three strawberries into their cups—the remainder ended up as "kid chow."

They did pick about one cup of strawberries so I saved them for breakfast. Man, what a meal it was! We had brook trout, strawberry pancakes with strawberry syrup, and hot Tang. That was a breakfast that fancy restaurants couldn't have bettered. There were no leftovers. Needless to say, the boys took all the credit for the breakfast and thought the camping trip was just great. They have enjoyed camping ever since.

Wild strawberries can help a hiking trip along. They are fairly common from the foothills to near the timberline. While seldom abundant enough for more than a cup or so, they sure perk up a weary hiker who has been climbing up switchback trails for hours. It sure is nice to take the pack off for a breather, look around, and find a patch of strawberries. One handful of the sweet, tasty berries gives a person a lift and renewed energy to tackle the trail ahead.

Strawberries are the only non-woody plant covered in this book. They belong to the genus *Fragaria* and are low-growing plants that send out runners. The plants seldom exceed four inches in height. The leaves are palmately compound with three leaflets originating at the same point on the stem. The leaflets are green on top with a light green underside. The leaf stems originate at the base of the plant.

Strawberries like sunshine so you'll not find them growing in the middle of a dense forest. You will find them in parks and clearings, at forest edges, on floodplains, and on open hillsides.

Strawberries bloom May through July depending on the elevation. The blossoms have five white petals surrounding a yellow center. The blossom is slightly smaller than a dime in size. Wild strawberries take about one month to mature. They look like small domestic strawberries when they ripen. (And well they should, for our domestic varieties were derived by crossing the various wild species). The size of the berry ranges from pea-sized to about the size of a small domestic grape.

Once in a while you will find a wild strawberry patch where you can pick enough to make jam (two quarts), but I would put your chances at one in ten for such a find. Your chances for finding a cup of wild strawberries is about one in three.

Wild strawberries are quite soft, so if you do get into a good patch of them, don't stack them up more than two inches high in your bucket. If you stack them high, you'll have a bucket of mush by the time you have a full bucket. 🍓

Recipes

NO-BAKE BERRY CHEESECAKE

1/2 a 10-ounce package shortbread cookies, finely crushed
4 tablespoons butter or margarine, softened
1 envelope unflavored gelatin
water
2 8-ounce packages cream cheese, softened
1/2 cup sugar
2 eggs, separated, at room temperature
1 1/2 teapoons lemon juice
1 teaspoon grated lemon rind
1/2 teaspoon vanilla extract
2 pints wild strawberries
1 pint huckleberries

1. In 10'' x 2'' springform pan, mix by hand crushed cookies and butter or margarine; press onto bottom of pan; set aside.

2. In small bowl, mix gelatin with 1/4 cup cold water; let gelatin stand 5 minutes to soften. Add 3/4 cup VERY HOT tap water to mixture and stir until gelatin is completely dissolved—about 3 minutes.

3. In large bowl, using a mixer at low speed, beat cream cheese, sugar, egg yolks, lemon juice, grated lemon rind, and vanilla until mixed; gradually beat in gelatin mixture. Increase speed to medium; beat until mixture is very smooth, scrape bowl often with rubber spatula.

4. In small bowl, using a mixer at high speed, beat egg whites until stiff peaks form. With rubber spatula or wire whisk, fold egg whites into cheese mixture; spoon mixture over crust in pan; cover pan with plastic wrap and refrigerate till firm—about 3 hours.

To serve, remove side of pan from cheesecake. Arrange berries on cake.

Judy Faurot, Sheridan, Wyoming

WILD STRAWBERRY SLUMP*

Slump Batter:
Cream together 3 tablespoons butter or margarine and 4 tablespoons sugar.
Add 1/2 cup milk and blend.
Mix 1 1/2 cups flour, 1 1/2 teaspoons baking powder, 1/4 teaspoon salt.
Add dry ingredients to wet, mix. Drop batter by spoonfuls into Bubbling
Berries. Cover and cook for 10 minutes.

Bubbling berries:
4 cups wild strawberries
1 1/2 cups sugar
3 tablespoons cornstarch
1 1/2 cups water

Bring to a boil in a heavy saucepan. Cinnamon or nutmeg may be added
to taste.

*Blueberries, huckleberries, blackberries, raspberries can also be used.

Charlotte Heron, Missoula, Montana

WILD STRAWBERRY FREEZER JAM

3 1/4 cups wild strawberries
1/4 cup lemon juice
1 package MCP pectin
1 cup light corn syrup
4 1/2 cups sugar

Wash, stem, and crush berries. Measure the level cups crushed berries and lemon juice in 4-quart kettle and stir well. Sift in slowly 1 package MCP pectin, stirring vigorously. Set aside 30 minutes, stirring occasionally. Add 1 cup light corn syrup. Mix well. (Syrup keeps sugar crystals from forming.)

Measure 4 1/2 level cups sugar into a dry bowl. Gradually stir this into the crushed fruit. Warming to 100 degrees F. (or temperature you would use for baby's milk) will hasten sugar dissolving. No hotter, please! When sugar is dissolved, jam is ready to eat.

Keep in jelly glasses or suitable freezer containers with tight lids. Store in freezer. If jam is to be used in 2 or 3 weeks, it may be stored in refrigerator.

MCP Pectin, Anaheim, California,
Submitted by Charlotte Heron, Missoula, Montana.
"I use the freezer jam recipe . . . wild strawberries are so fragrant, I hate to cook them."

STRAWBERRY CREME PIE

Pie shell:
1 cup flour
1/2 teaspoon salt
1/2 cup shortening
enough water to make dough

Bake pie shell at 425 degrees approximately 10 minutes or until done.

Filling:
at least 1 pint strawberries
1 cup sugar
6 tablespoons cornstarch
1/2 teaspoon salt
2 1/2 cups milk
2 beaten eggs
1/2 teaspoon vanilla

Mix sugar, cornstarch, salt, milk. Cook over medium heat, stirring constantly so it doesn't burn on bottom (can use a double boiler). When it thickens, add small amount of mix to eggs, then stir into mix. Remove from heat, add vanilla. Cool.

Line baked pie shell with strawberries. Pour filling over berries. Add more berries on top (in a decorative pattern if you wish). Chill.

Charlotte Heron, Missoula, Montana

FROSTY STRAWBERRY SQUARES

1 cup flour
1/4 cup brown sugar
1/2 cup walnuts
1/2 cup butter or margarine, melted

Stir together; spread evenly in shallow baking pan. Bake in 350 degree oven 20 minutes, stirring occasionally. Sprinkle 2/3 of the crumbs in a 12" x 9" x 2" baking pan.

2 egg whites
1 cup granulated sugar
2 cups fresh wild strawberries
2 tablespoons lemon juice
1 cup whipping cream, whipped

Combine egg whites, sugar, berries, and lemon juice in large bowl; with electric beater, beat at high speed into stiff peaks, about 10 minutes. Fold in whipped cream, spoon over the crumbs; top with remaining crumbs. Freeze 6 hours or overnight, cut in squares to serve. If desired, trim with whole strawberries.

Mary Lohuis, Jackson, Wyoming

OLD-FASHIONED SUNSHINE STRAWBERRY JAM

1 1/2 pounds (1 1/2 quarts) ripe strawberries
4 cups sugar
juice of 1 lemon (3 tablespoons)

Wash and hull strawberries. Put into a kettle with sugar and lemon juice and heat slowly to boiling point. Cook rapidly about 10 minutes. Pour berry mixture into 1/2-inch deep platters or shallow containers. Cover with a sheet of glass or plastic wrap, propping the cover up to leave about 1 to 2 inches between it and the pan to allow for evaporation. Set outside in the sun. As the fruit cooks in the sun, turn it with a spatula 2 or 3 times during the day. If the sun is not hot enough, or a wind comes up during the day, the jam can take 2 or 3 days before it is ready. When it has thickened enough, pour into hot, sterilized jars and seal. Process in boiling water bath for ten minutes. Yield: Approximately four 8-ounce jars.

Cel Hope, Sheridan, Wyoming

Calendar

	Blossoms*	Ripens*
Black Currant	late April to mid-May	late July—August
Buffaloberry	late April to mid-May	mid-July—August
Chokecherry	May	August
Elderberry	June	mid-September
Gooseberry	late April to mid-May	late July—August
Hawthorn	mid-May to early June	late August—Sept.
Huckleberry	late May—June	August—mid-Sept.
Low Bush Blueberry	late May—June	late July—mid-Sept.
Oregon Grape	mid-April—mid-May	September
Serviceberry	late April—mid-May	mid-July—mid-Aug.
Wild Grape	mid- to late May	September
Wild Plum	late April to mid-May	September
Wild Raspberry	mid-June—mid-July	mid-Aug.—mid-Sept.
Wild Rose	June	September
Wild Strawberry	June—mid-July	July—mid-August

* A plant will blossom earlier at lower altitudes and later at high altitudes. The same holds true for when the berries or fruits ripen.

Glossary

Alternate: refers to leaf arrangement. There is only one leaf per node on alternating sides of the stem.

Biennial: completes life cycle in two years.

Blush or Bloom: a whitish, powdery covering of the fruit, berry, leaf, or twig.

Cane: a pithy stem found in raspberry or elderberry.

Cyme: a flat or nearly flat-topped flower cluster.

Dioecious: having the female flowers on one plant and the male flowers on another. Buffaloberry is dioecious.

Glabrous: completely smooth without any hairs or bristles.

Hip: a fleshy, cup-like receptacle that is the rose fruit.

Inferior Ovary: the flower parts arise from the top of the ovary.

Lanceolate: Lance-like. The leaf is several times longer than it is wide; the base is broadest, tapering to a point.

Lenticels: breathing pores in the bark that resemble warts or light-colored spots.

Mesic: mid-range of an environmental variable, such as moisture.

Opposite: there is a leaf on each side of a stem at each node. The leaves are opposite one another. This contrasts to alternate leaf arrangement, where there is only one leaf at the node.

Palmately Compound Leaf: the leaflets arise from a central point.

Pinnately Compound Leaf: the leaflets arise along a central stem.

Pome: Fleshy fruit from an inferior ovary. Example: apple.

Raceme: An inflorescence or cluster of flowers along one main stem.

Render: a cooking term meaning to extract juice or pulp.

Sepal: part of a flower which is situated beneath the petals. The sepals are often green-colored.

Serrate: having small teeth.

Sheet Test: a cooking term that refers to the jellying point. Take a spoonful of hot jelly from the kettle and cool a minute. Holding the spoon at least a foot above the kettle, tip the spoon so the jelly runs back into the kettle. If the liquid runs together at the edge and ''sheets'' off the spoon, the jelly is ready.

Stamen: the male reproductive part of the plant that produces pollen.

Talus slope: the accumulated rocks below a peak or knob.

Tendril: An outgrowth at the end of a leaf or stem node used for clinging or climbing.

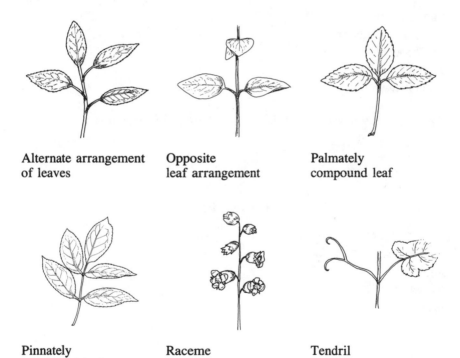

Alternate arrangement
of leaves

Opposite
leaf arrangement

Palmately
compound leaf

Pinnately
compound leaf
Elderberry

Raceme

Tendril

Berry Good Hints

Snack Berries*

Huckleberry
Low Bush Blueberry
Serviceberry
Wild Raspberry
Wild Strawberry

*These are berries you can eat as is without any preparation.

Pie Berries

Black Currant
Chokecherry
Elderberry
Gooseberry
Huckleberry
Low Bush Blueberry
Serviceberry
Wild Plum
Wild Strawberry

Versatile Recipes*

* You can substitute one or more berries in these recipes.

Camp Recipes*

* These are recipes you can make over a camp stove or campfire.

Index

Recipe Index

Bibliography

Ball Corp. *Ball Blue Book*, Muncie, Indiana; Ball Corporation, 1989.

Bindl, Betty and Freida Park. *The Trout Creek Huckleberry Cookbook.* Trout Creek, Montana, 1984.

Brannon, Dave and Nancy. *Feasting in the Forest.* Cody, Wyoming: Dave and Nancy Brannon, Publishers, 1989.

Butterfield, Margaret and Charles. *Preserving Wyoming's Wild Berries & Fruit.* Laramie, Wyoming: Agricultural Extension Service B-735, 1981.

Holmgren, Arthur H. *Handbook of the Vascular Plants of the Northern Wasatch.* Palo Alto, California: The National Press, 1965.

Jensen, Earl R. *Flowers of Wyoming's Big Horn Montains and Big Horn Basin.* Basin, Wyoming: Basin Republican Rustler Printing, 1987.

Kerr Glass Manufacturing Corp. *Kerr Home Canning and Freezing Book.* Los Angeles, California: Consumer Products Division. 1982.

MCP Foods. *Make a Honey of a Jam with Recipes from MCP Foods.* Anaheim, California: MCP Foods.

MCP Foods. *Special Recipes from MCP Foods.* Anaheim, California: MCP Foods.

Michels, Joyce. *The Billings Gazette Cookbook.* Billings, Montana: The Billings Gazette, September 25, 1983.

Montana Department of Fish, Wildlife, and Parks. *Savoring the Wild.* Helena, Montana: Falcon Press, 1989.

Muenscher, Walter Conrad. *Poisonous Plants of the United States*. New York, New York: Collier Books, 1975.

Nelson, Ruth Ashton. *Handbook of Rocky Mountain Plants*. Estes Park, Colorado: Skyland Publishers, 1977.

Rydberg, P.A. *Flora of the Rocky Mountains and Adjacent Plains*. New York, New York: Hafner Publishing Company, 1969.

Swan Lake Women's Club. *Huckleberry Recipes Compiled for your Pleasure*. Swan Lake, Montana.

Van Bruggen, Theodore. *Wildflowers, Grasses & Other Plants of the Northern Plains and Black Hills*. Rapid City, South Dakota: Badlands Natural History Association.

Weber, William A. *Rocky Mountain Flora*. Boulder, Colorado: Colorado Associated University Press, 1976.